WHALE SONG

'Never confuse credentials with qualifications. Andrew Stevenson may lack credentials, but he's thoroughly qualified (and I'll take qualifications every time). His great achievement is to show how someone with energy and persistence can make a solid contribution to knowledge about any animal that lives in their back yard – even when that animal is the humpback whale and that back yard, the North Atlantic. In four years Andrew has identified more than twice as many humpback whales around Bermuda as we scientists identified in those waters in the previous 40 years.'

Dr Roger Payne, founder of the Ocean Alliance, who discovered in 1967, with Scott McVay, that humpback whales sing songs

---

'A terrific book, a real pleasure to read. Its science is sound, its speculations appropriate, and its spirit is generous and uplifting. It finds the place where the curiosity of people and whales meet. An engaging and informative result from hard and persistent work.'

Dr Steven Katona, Managing Director, Ocean Health Index, Conservation International; Adjunct Senior Scientist, New England Aquarium

---

'Andrew Stevenson's passion for the Humpback whale has made of him a profound student of these exquisite leviathans. Meticulously researched and expertly documented by an engaging writer and gifted photographer, this book conveys the exhilarating experiences of Bermuda's leading expert and defender of whales. Some books are more than books, they are treasures to be cherished, and this is one of them.'

Captain Paul Watson, Sea Shepherd Conservation Society

'Andrew's whale adventure is a rare and wonderful leap into the world of the whale.
His eye-to-eye meeting with a humpback whale is the stuff that changes lives
and to follow him in his search for meaning harkens back to the passions
that started the "Save the Whales" movements forty years ago.
It is very refreshing that those fires still burn.'
Flip Nicklin, whale researcher and National Geographic photographer

---

'Courageous, scientific and inspirational. Andrew has skilfully woven a riveting personal account of his experiences with these incredible creatures which makes readers feel as though they are in the water right next to him. Whale Song *highlights just how much we have yet to learn about the world's oceans and the species found in them,*
and the profound and irreversible impacts we humans are having on them.'
Dr Greg Stone, Chief Scientist for Oceans,
Senior Vice-President Marine Conservation, Conservation International

---

'Most impressive is the way in which Andrew has successfully combined sometimes conflicting attributes – personal dedication and emotional commitment with the objectivity and impartiality required for observational science. Reading his "voyage of discovery" into the seldom-seen world of these great travellers makes us all long for a passion that takes over and propels us to great achievements.'
Dr Steven Gittings, Science Director, National Marine Sanctuary Program,
National Oceanic and Atmospheric Administration

'Elegant writing that is easy to read and very interesting throughout. I learned a lot. I have been aware of bits and pieces of what Andrew has been doing, but have not had the whole picture. Now I do. Great stuff.'
Dr Jim Darling, humpback whale researcher for over thirty years and author of *Humpbacks, Unveiling the Mysteries*

'This entrancing story is a delightful and riveting read. A real-life story of a father and daughter's quest to learn about whales. Andrew's powerful, lyrical writing sweeps you along – you won't be able to put Whale Song *down.'*
Chris Palmer, wildlife film producer, Distinguished Film Producer in Residence and Director of the Center for Environmental Filmmaking at American University, and author of *Shooting in the Wild: An Insider's Account of Making Movies in the Animal Kingdom*

'Andrew's Magical Whale encounter would make any whale biologist jealous!'
Hilary Moors, PhD candidate, Marine Mammal Acoustics and Behavior, Whitehead Lab, Dalhousie University

---

'Will illuminate the imagination of its readers, scientists and non-scientists alike. Andrew spreads his knowledge, wisdom and excitement about whales to people with an audacity unlike others, despite working in the exposed North Atlantic Ocean!'
Tara Stevens, PhD candidate in the Cognitive and Behavioural Ecology Programme at Memorial University of Newfoundland

---

'What a great book! Andrew has a delightful expressive way.'
Rosemary E. Seton, Whale Researcher, North Atlantic Humpback Whale Catalogue; Marine Mammal Stranding Coordinator, Allied Whale Marine Mammal Research Lab, College of the Atlantic

---

'Andrew has caught them on the fly – in the Bermuda Triangle, of all places! That's no mean feat, and the outcome is a tantalizing glimpse of the private lives of humpbacks. Let's have more of it!'
Dr Wolfgang Sterrer, former Curator, Bermuda Natural History Museum

---

'A book of discovery about how one can grow and achieve difficult goals. Andrew shows us how much wonder there is to be revealed in our oceans, when we take the time to look and ask the right questions.'
Dr Robbie Smith, Curator, Bermuda Aquarium, Museum and Zoo

'This is a book about persistence, passion and discovery. Through his sheer tenacity and devotion to his daughter, Andrew has provided a wealth of information about Atlantic humpback whales for all of us. We should all thank Elsa for asking the question "why?"
Dr Ian Walker, Principal Curator, Bermuda Aquarium, Museum and Zoo

---

WHALE SONG

Journeys into the Secret Lives of the North Atlantic Humpbacks

ANDREW STEVENSON

CONSTABLE • LONDON

CONSTABLE & ROBINSON LTD
55–56 RUSSELL SQUARE
LONDON WC1B 4HP
WWW.CONSTABLEROBINSON.COM

FIRST PUBLISHED IN THE UK BY CONSTABLE,
AN IMPRINT OF CONSTABLE & ROBINSON LTD, 2011

PUBLISHED IN COLLABORATION WITH:
THE HUMPBACK WHALE RESEARCH PROJECT, BERMUDA
16 SKYLINE ROAD
SMITHS FL08
BERMUDA
WWW.WHALESBERMUDA.COM
TEL. 1 441 77-SPOUT (1-441-777-7688) SPOUT@LOGIC.BM

A COPY OF THE BRITISH LIBRARY CATALOGUING IN PUBLICATION DATA IS AVAILABLE FROM THE BRITISH LIBRARY.

ISBN: 978-1-84901-617-9

PRINTED AND BOUND IN CHINA

10 9 8 7 6 5 4 3 2 1

In memory of

Deborah Barbour Butterfield

Raised in St John, New Brunswick, she knew
the waters of the North Atlantic humpbacks' journey,
from their breeding grounds in the Caribbean
and their mid-ocean migratory route past
her home in Bermuda to their feeding
grounds in the Bay of Fundy.

"It is because whales are such grand and glowing creatures that their destruction degrades us so. It will confound our descendants. We were the generation that searched Mars for the most tenuous evidence of life but couldn't rouse enough moral outrage to stop the destruction of the grandest manifestations of life here on earth."

Dr Roger Payne

CONTENTS

WHALE SONG

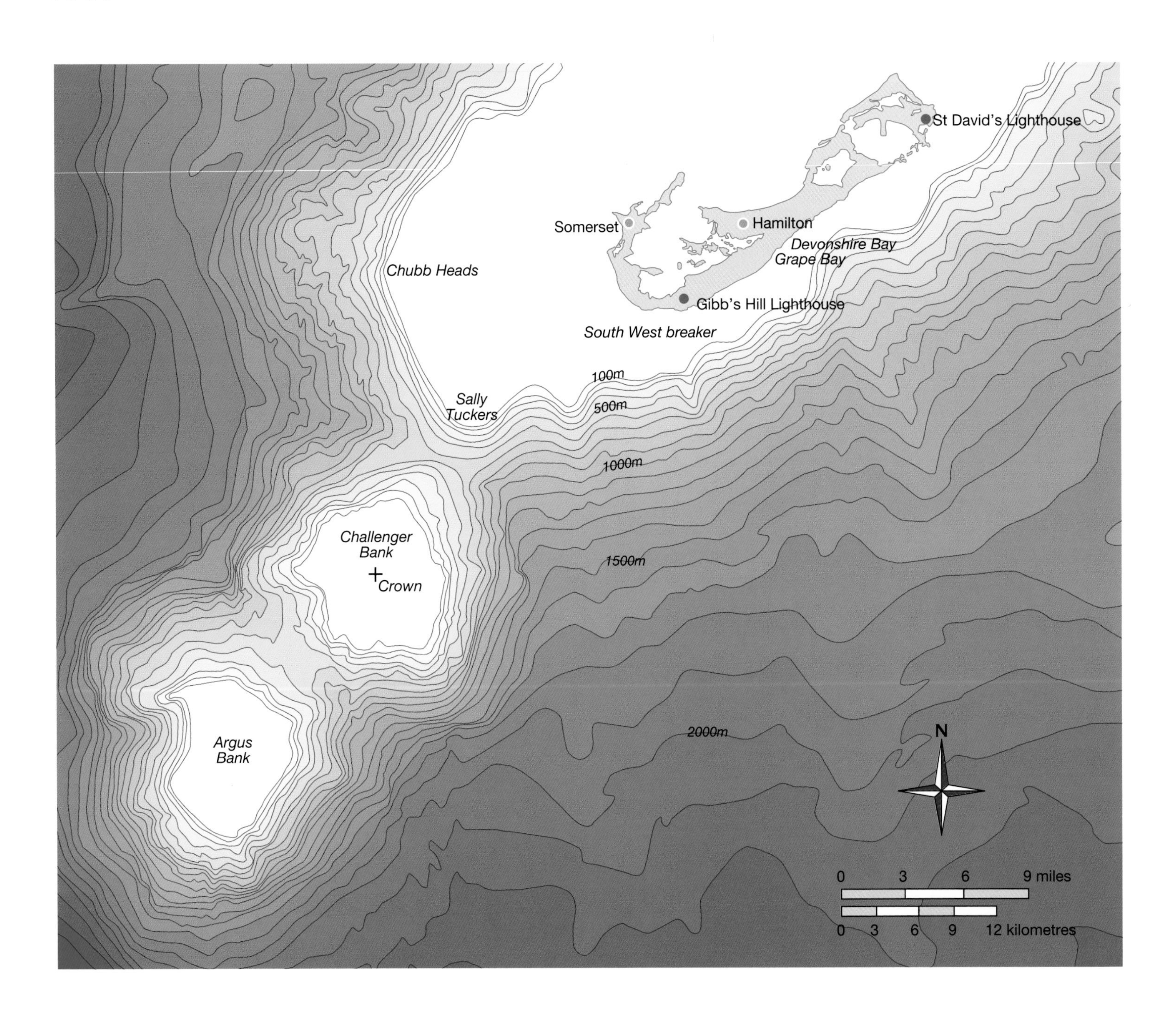

St David's Lighthouse
Somerset
Hamilton
Devonshire Bay
Grape Bay
Chubb Heads
Gibb's Hill Lighthouse
South West breaker
100m
500m
Sally
Tuckers
1000m
Challenger
Bank
Crown
1500m
Argus
Bank
2000m
N
0
3
6
9 miles
0
3
6
9
12 kilometres

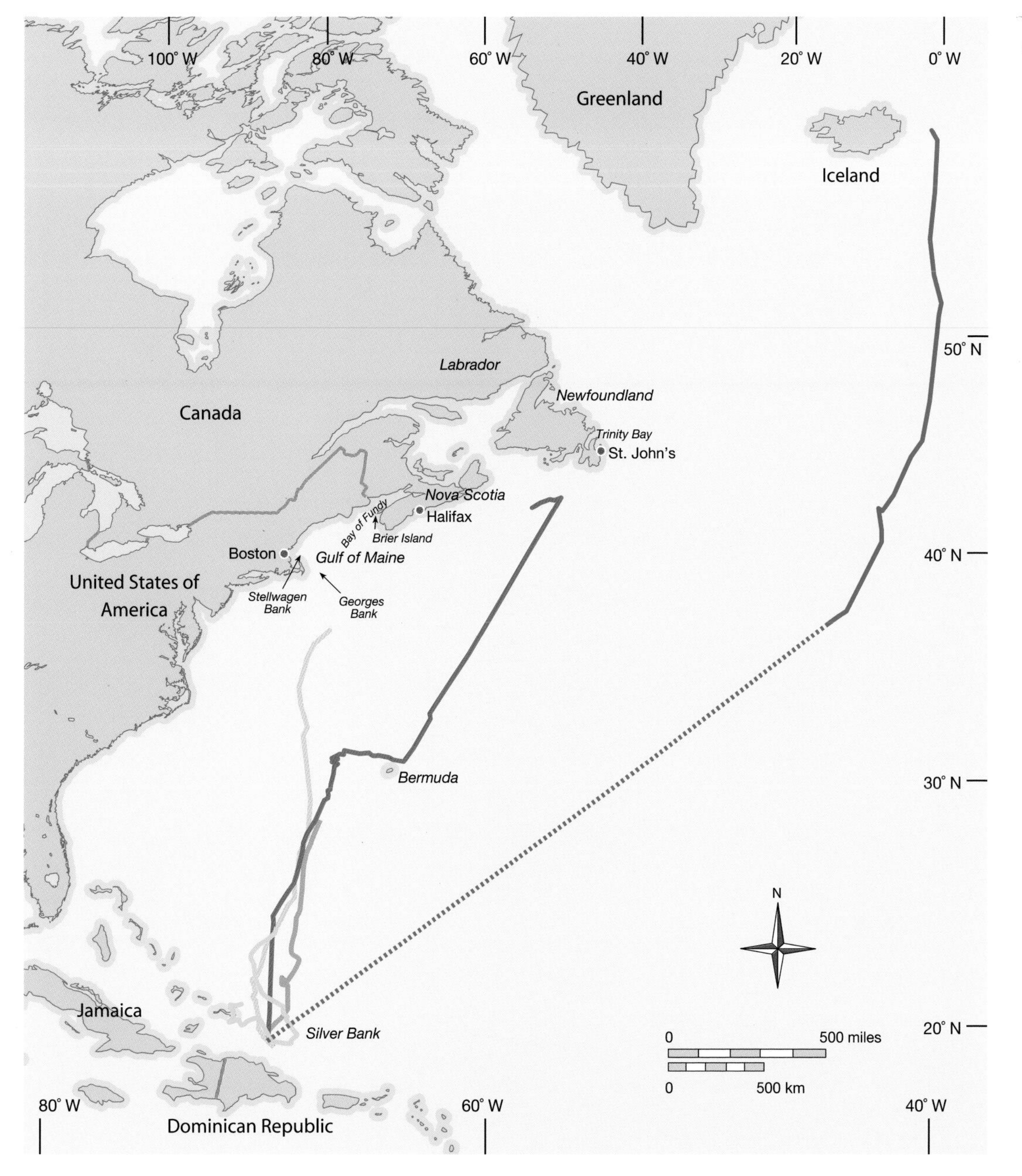

(left page) Map of Bermuda showing its shallow platform extending several miles to the north and southwest to an area we call Sally Tuckers. The two seamounts, Challenger Bank and Argus Bank lie some fifteen and twenty-five miles offshore to the southwest.

(this page) Map showing the tracks of satellite-tagged whales as they migrate northwards from the Silver Bank to their feeding grounds up north. The red line shows the path the mother and calf took in diverting towards Bermuda. The broken blue line represents the presumed path of a whale with a tag that did not transmit until the whale was west of the Azores.

FOREWORD

For those expecting only a book about whales, this isn't it. Andrew Stevenson's *Whale Song* is far bigger, better and richer than that. It is at once an ambitious personal quest, a loving gift to a child, a careful scientific investigation, a meditation on persistence and a journey to the place where the curiosities of humans and whales meet.

It is the story of two species learning how to survive and thrive in a changing world, while we humans learn to do a better job of understanding and caring for our fellow travellers on the planet; and whales try to overcome new challenges that will confront them until we do.

Every spring, humpback whales migrate north through Bermuda's crystal blue waters, where the modern study of the species, especially their remarkable songs, began in the late 1960s. Ever since, scientists have tried to explain why the whales come to Bermuda and what they do there. In *Whale Song*, Andrew Stevenson answers some of those long-standing questions. Readers will enjoy his accounts of research cruises in Bermuda, as well as to the whales' feeding ranges in New England and Canada, and their breeding range in the West Indies. I'm confident that every reader will find his observations, experiences, questions and speculations as fascinating as I did.

Though it is still too early to tell whether the story will end happily for all, there is reason for optimism. Despite the many ways that human activities and carelessness impact the ocean, and despite the toll that they have taken on marine life, most populations of large whales, including the humpback whales featured in *Whale Song*, are increasing. Thanks to international treaties such as the International Whaling Convention and the Convention on International Trade in Endangered Species (a worldwide moratorium on commercial whale hunting that has been in force for twenty-five years) and protective legislation enacted by individual countries including Canada, Australia and New Zealand, populations have rebounded strongly from alarmingly low levels in the 1960s.

This success story is both a beacon of hope and a guide for ocean restoration. Effective legislation, supported by a strong public mandate, can restore the ocean and its populations to health. Whales have benefited from the efforts of countless citizens, scientists, diplomats and marine managers who rallied to their defence. Similar efforts can bring success for the rest of the ocean's life and all of us who depend on it. All we need is the personal and political will to bring that about. *Whale Song* will inspire everyone towards that goal.

Dr Steven Katona

Bar Harbor, Maine

"My hope is that this film and book will ignite the passion of a child
to discover and share the secrets of the North Atlantic humpback whales."

Andrew Stevenson

PROLOGUE

My journey with the whales began when Elsa was almost three years old. It was a lovely spring morning in Bermuda. We were walking along the beach at the bottom of our road looking for critters when a whale jumped out of the water on the other side of the coral reefs, not much more than a hundred yards from shore.

'What was that?' Elsa asked.

'A whale,' I replied.

'Why does it do that?' Elsa asked.

I didn't know.

There was not a lot I did know about whales. I'm not sure, when I think back, that I even knew what kind of whale it was that we had seen. Seeing a prehistoric-seeming animal weighing scores of tons cast itself out of its watery kingdom into our own domain, almost a stone's throw away, was an epiphany for me. Like Jacques Cousteau when he dived for the first time with a home-made mask, lifted his head partially out of the water and simultaneously saw the cars on Toulon harbour above and the underwater realm of the Mediterranean fish below, I saw this whale through a prism revealing two parallel worlds. Here was a species that had existed for millions of years in much the same form as I had just witnessed. Long before Europeans had colonized the New World, long before Bermuda was settled 400-plus years ago, long before *Homo sapiens* stood upright and tamed the world, these ancient mammalian cousins of ours inhabited the oceans. And for the most part, compared to the radical transformations made by humans on land, the depths and secrets of the planet's oceans remained largely unfathomed.

In order to answer Elsa's question, I began researching the humpbacks, the most charismatic and studied of all the large whales. Humpbacks have been investigated extensively over the last three or four decades in their breeding and feeding grounds close to shore, but there was little that marine scientists knew about their mid-ocean lives. Considering these were such large animals, creatures that matched and surpassed the dinosaurs in size, this insight seemed astonishing.

I was familiar enough with the geography of the North Pacific to know that there was no terra firma to provide a handy launching pad into the whales' migratory crossings between their breeding grounds in Hawaii and their feeding grounds off the west coast of Canada, Alaska and Russia. Once they migrated north from Hawaii they effectively disappeared until they reached their northern feeding grounds.

But the North Atlantic was different. Bermuda lay smack in the middle of the humpbacks' migratory crossings from their breeding grounds in the Caribbean to their northern feeding grounds in the eastern United States and Canada. This was no revelation – from the scientific papers I started to read, many well-known marine biologists had come to Bermuda at one point or another to study these whales.

Although Bermuda is better than an ocean-going ship as a platform from which to study whales, it is nevertheless in the middle of the ocean, exposed to the open North Atlantic's wind and waves. We are often buffeted by winter storms when the whales' migration reaches Bermuda in March and April. Searching the open ocean for a whale's blowing spout, or spotting the tell-tale silhouette of a humpback's back or raised flukes amidst wind-flung spray and heaving waves, is challenging. Compounding this drawback, the whales off Bermuda do not remain for some months as they do in the breeding and feeding grounds: they are in transit, apparently cruising right by us on their way north. For a marine scientist to come to Bermuda for three weeks only to get out on the water a couple of times because of bad weather is not the most productive use of time. Over the decades many leading marine scientists gave up researching the humpbacks' mid-ocean social behaviour based out of Bermuda simply because it was too difficult.

But for someone living in Bermuda, this was an opportunity to be at the leading edge of scientific discovery. Making this possibility even more enticing is the immense physical size of the subject of this potential exploration. Whales are not tiny organisms that require examination under a microscope – they are larger-than-life characters, impressive not only in size, but in intelligence as well.

The prospect of finding out about the humpbacks' lives in an environment that had never been studied before – underwater in the middle of their migratory crossings in the heart of the ocean – was, to say the least, compelling. Belatedly, I realized I didn't have to travel to the far corners of the earth to find adventure. Instead of exploring the Himalayan mountains as I had been doing for the last couple of decades, I'd be sitting on the top of a mid-ocean mountain and immersing myself, literally, into the complex underwater lives of the humpbacks. Potential cutting-edge exploration existed on the doorstep of our tiny Bermuda cottage.

I knew myself well enough that, if I embarked on a project, I would not let it go until I had achieved what I had set out to achieve. 'A terrier with a bone,' as one of my volunteers described me. A tenacious personality can be an asset, but it can also lead to disaster if the undertaking is something impossible to achieve.

Whale-watching boat captains here in Bermuda told me that if they were lucky, once a year, a curious whale, usually a calf, would 'mug' their boat. I calculated that if I went out on the water the full two months that the humpbacks were reportedly migrating by, I might have one or two chances to get in the water and film a calf. If I went out at every opportunity over a three-year period, I might have three or four underwater encounters with whales that might give me good enough footage to use in a short documentary film for schoolchildren in Bermuda. It would not be easy, but it might be possible – simply because I live here and could maximize the calm-weather days when I went out on the water.

I did not have any experience taking photographs underwater, never mind using a video camera. I didn't even know whether it was possible for an amateur to attempt a project like this. Perhaps I needed years of training in filmmaking and a background as a marine biologist. I cold-called Flip Nicklin in Hawaii, regarded as the premier whale photographer of the world, and a *National Geographic*-contributing photographer with more than 5,500 dives under his belt. His ability to free-dive to depths of up to 90 feet (27 metres) allows him to swim near enough to record whale behaviour without interrupting the whale.

Flip was not only very encouraging, he was also very generous with his advice to a complete stranger. His conclusion was very simple: 'With the advances in high-definition video cameras and underwater housings, if you are comfortable in the water, you'll have no problem.'

I am comfortable in the water. I train every other day in a pool and swim open water races. I did not know if I could swim down to 90 feet and make it back up to the surface again with a bulky camera, but I was encouraged enough to start researching high-definition video cameras and underwater housings.

The equipment I would need would have cost hundreds of thousands of dollars a few years ago but advances in technology meant that I could get what I needed and start filming and editing for as little as US$65,000.

The next question was funding. No one relishes fundraising, even if it does finance a project that has tremendous personal appeal. It took just a few calls to Bermudians whom I knew were interested in the ocean and I had my first donations. I can only assume their willingness to support the project had more to do with the appeal of whales than any confidence in a late-in-life, wannabe marine biologist-filmmaker with no background in either field.

With the first donated cheques deposited into the project account, I got off the fence and jumped in at the deep end, literally. There was no turning back. I was committed to making an underwater film about North Atlantic humpbacks. For the next three years I would do everything possible to obtain the first high-definition underwater footage of humpback whales on a mid-ocean seamount (subterranean mountain) in the middle of their migratory crossings.

By now Elsa was more than three-and-a-half years old and a willing accomplice.

CHAPTER ONE

MAGICAL WHALE

That first season in 2007 trying to film whales was disastrous. Despite 300 hours trying to find humpbacks and get into the water with them, I didn't have a single minute of usable footage. After six weeks attempting to film whales underwater, I had carpal tunnel syndrome, causing pain and weakness in my wrists, from lugging the 55-pound (25 kg) underwater housing from our home to the car, the car to the punt, the punt into the boat and vice versa. My body was bruised and worn out from back-to-back, nine-hour days standing on the swaying tuna tower of a small single-engine boat. My eyes were red from searching for whales. I was wind-burned and sunburned. I had found the whales, but unlike whale-watching from the surface, getting into the water to film them is much more difficult. Visibility underwater is not the same as it is above. The whales have to be extraordinarily close and, because they are in their element and can swim ten times faster than I can, they have to approach me. Swimming with a whale has to be on the whale's terms. It is up to the whale to initiate close underwater contact.

The lack of underwater footage wasn't for lack of trying, however. I had gone out every day that weather permitted, and on some days when it hadn't. Much of the winter and spring weather in 2007 was atrocious and getting in and out of Devonshire Bay on South Shore, where I kept the borrowed boat, was often a terrifying ordeal. Sometimes it felt as if I was in a lifeboat rocketing through the waves that battered the shore at the entrance to the bay. Navigating through the labyrinth of reefs protecting the cove added another challenging dimension. Even the local fishermen who have been boating out of Devonshire Bay for decades thought I was crazy to attempt to leave the bay in the wind and wave conditions I went out in.

After two months of constant effort I was frustrated, tired and empty-handed. I kept telling myself that if it was easy to film whales underwater in the middle of the ocean, everyone would be doing it. I had known it would be an ambitious undertaking which is why I had given myself three years to make a thirty-minute documentary for Bermuda schoolchildren. At the end of that first season, I reminded myself that I still had another two seasons to go as I kept heading out, day after day, despite the local fishermen's opinions that the migrating whales had already gone.

Then the persistence paid off. It was almost May 2007. We trolled once again along South Shore westwards of Devonshire Bay and continued past Sally Tuckers, the south-west corner of the Bermuda platform. We crossed the 4,000-foot (1,220 m) deep canyon separating the Bermuda platform from a seamount called Challenger Bank, 15 miles (24 km) offshore. Bermuda was now out of sight and we were effectively cast loose in the middle of the Atlantic. For a small single-engine outboard, this was a long way out into the open ocean. By midday we hadn't seen a thing. We stopped, took a lunch break and I showed my two volunteer crew for the day, Kelly Winfield and Kevin Horsfield, how to help me change the video clips on the underwater video camera. I also showed them

The first season we found whales one out of every three trips we went looking for them. Often we had pods of North Atlantic bottlenose dolphins come to us to ride the pressure wave on the bow of the boat.

how to use another video camera on the boat, should the occasion arise. That, as it turned out, was a good thing.

Soon after our lunch break a radio call from a fishing boat reported whales in the north-west quadrant of Challenger Bank. We started the engine and looked for the whales but instead came across a pod of dolphins. It was a mirror-like calm day and I took advantage of the conditions to film the dolphins from the bowsprit of our little boat, holding the camera inches above the ocean surface as the dolphins rode the boat's bow wave. I put on my wetsuit and slid into the water with the underwater camera. Not shy when I was filming them onboard the boat, the dolphins were leery of me when I was in the water. I noticed a change in their body language and their clicking sounds. I glanced down and saw a dark shape immediately beneath me.

It was a humpback whale. The whale swam beside me, looked me in the eye, and stared. It was only feet away.

After weeks of searching for whales, this whale had found me.

The adult, full-grown humpback performed intricate ballet movements around me, under me and beside me. At times it almost touched me. I did not have a sense of fear, although I did feel intimidated as the 45-foot (14 m) animal repeatedly came directly at me and then dived, the massive 12-foot (4 m) wide flukes passing within arm's reach underneath me. A deliberate sideswipe of its huge tail would render me pink goo in a wetsuit. And yet this huge animal was incredibly gentle. It always knew exactly where I was in relationship to any part of its body. The delicate tips of its long 16-foot (5 m) flippers flexed – like the fingers of a human hand – as they reached out and almost stroked my face. Often it lay quietly on its back underneath me as if embracing me, his flippers extending along either side of me.

The massive fluke of its tail flicked within inches of me. We were so close to each other that when the whale dived, the vortex of water sucked me down behind the 45-ton cetacean (Cetacea is the order of whales, dolphins and porpoises). Over and over again it placed his gigantic head a foot or so underneath me and remained motionless, as if listening to my heartbeat. But it never made a sound. I could easily have reached out and patted this gargantuan animal on its head as it hung in the water beside me, its tennis-ball-sized gray eye peering intently at me.

The experience was unreal. I did not feel scared. The whale had approached me when I was already in the water. The decision was his, not mine. I was a mute bystander to what must have been a unique experience for both of us. Were it not for the fact that I had an underwater video camera documenting this, and another video camera recording the encounter from the boat, this might have been the ultimate fisherman's tall tale of a close encounter with a giant whale.

This was not a curious calf swimming by with its indulgent mother though. This was an adult male apparently attempting to make contact.

It was a powerful one-on-one experience, the whale totally focused on my presence. Several times his fins or flukes pummelled the water within inches of me, as if testing my resolve, but strangely, I still didn't feel threatened. I was intimidated, yes, most definitely. But I knew this humpback understood I wasn't a danger to him.

The only time I felt anxious was when I was about a hundred feet from the boat and another whale approached. The two whales scuffled and I was caught in the middle. All I could see was white foam from the thrashing of their 12-foot (4 m) flukes and 16-foot (5 m) pectoral fins. I was sucked down in the maelstrom of water, but neither whale hit me. Eventually the other whale swam off, but not before my crew members had photographed its uplifted tail.

During the course of the two hours that I was in the water, fishing boats approached us, aware that something was going on. On both occasions, despite the boats keeping a respectful distance, the whale swam off. But when the spectators motored away, this magical whale returned to us like a pet dog. When I had to change the high-definition clips in the camera and climbed into the boat

(previous page) At the end of the season the fishermen told me all the whales had already gone. Then I had my first in-the-water, close-up experience with a humpback -- a two-hour, intimate encounter with Magical Whale.

The relaxed encounter with Magical Whale changed when another whale approached. The two whales threatened each other with their pectoral fins and flukes and I disappeared in the middle of their displays for dominance.

(overleaf) When I climbed into the boat to change the memory cards in the high-definition video camera, Magical Whale came alongside the boat and trumpeted through his blow hole and splashed us with his fluke. But once I slipped back into the water he returned as calmly as a pet dog, only this was a wild animal in its own element, an animal almost twice the length of our 25-foot boat.

the whale slapped the water beside us with its tail and pectoral fins. It trumpeted right beside the boat like an enraged elephant, the spray from his blowholes saturating us with vapour.

The whale was almost twice the length of our 25-foot (7.6 m) boat. Surfacing beside us, it seemed to have the dimensions of a submarine. During the time I was in the boat, as if annoyed that I was no longer in the water with it, the whale continued to trumpet and whack the water with its pectoral fins and flukes, just missing the sides of the boat.

When I finished reloading the camera, I slipped off the stern back into the ocean. The whale swam directly towards me until we were once again feet apart. It continued its serenade, effectively seducing me.

There were times when Magical Whale (that's what we decided to call him some days later) wasn't so embracing in attitude. Sometimes he seemed to test my tenacity. He could have swatted me as easily as I might swat a fly or step on an ant. And, considering how humans slaughtered these whales to the brink of extinction, there was every reason for this whale to take retribution. Its mother and father, its brothers and sisters, its offspring, could easily have fallen victim to the harpoons before the worldwide moratorium on whaling in 1986 ended the massacre.

There were occasions when this whale did flick his tail at my tiny body, creating a splash and surge of water that overwhelmed me. There were times when it launched its 45 tons directly at me along the surface, only to drop below, his massive flukes rising and then plunging vertically downwards in front of the camera. At other times he lunged at me only to deftly duck underneath so that the entirety of his 45-foot (14 m) length passed not more than a foot or so beneath me. Each time this happened his lethal tail remained motionless as he slipped harmlessly by.

I noticed a pattern. The further from the boat I was, the more physical he became. The closer I was to the boat, the more respectful he was, as if the boat was my mother and would protect me. At one point I was separated from the boat. Magical Whale began one of his passes at me and I swam across his intended path so that I wouldn't be cut off from the boat. This was a miscalculation. By cutting across his path we were on a collision course. There wasn't enough room for the whale to raise its flukes and dive under me. Instead, he arched his back, dragging his massive flukes like a sea anchor while spreading his pectoral fins wide and angling them vertically like the flaps on a jumbo jet. He came to an immediate stop. My underwater footage reveals this braking action as his fins back-pedal to arrest his momentum. The above-water footage by volunteer crewmember Kevin Horsfield in the boat is even more dramatic.

Without a 120-degree wide-angle dome-port, much of this underwater close-up action would have been meaningless. The extreme wide-angle optics of the lens captured most of the whale in the viewfinder, even if it did make the whale seem further away from me than it was in reality. This distant wide-angle perspective contrasts with the above-water footage. The underwater footage is often dreamlike and peaceful while the above-water footage is alarming, showing how close we were and giving a striking demonstration of the differences in scale between a human and a full-grown humpback.

When a fishing boat trailing several fishing lines started circling persistently around us, Magical Whale disappeared, this time for good. The entire encounter lasted two hours and yet it seemed like seconds. It was an intense, profound and eventually disturbing experience. For a week afterwards, I found it difficult to sleep. Over and over again I kept asking myself, 'What was he thinking?'

Side by side, eye to eye, staring intently into the window of the soul of a highly intelligent wild animal that dwarfed me in size was a both a humbling and powerful experience. I was at a complete loss for words to describe what I felt.

I couldn't understand why he had swum to me or why he had behaved the way he did. I kept asking myself what was going on in the brain of this animal. I wondered what he was thinking when he held his head up out of the water so that our eyes were at the same level while he looked at me so intensely.

Magical Whale effectively seduced me with his mesmeric dancing. There was no doubt in my mind that he was trying to communicate. The 120-degree, super-wide Gates port on the underwater housing provided a complete rectilinear image of Magical Whale from mere feet away.

At times Magical Whale rolled onto his side and extended his pectoral fins on either side of me, the tips of the 16-foot fins flexing as if he were going to caress me.

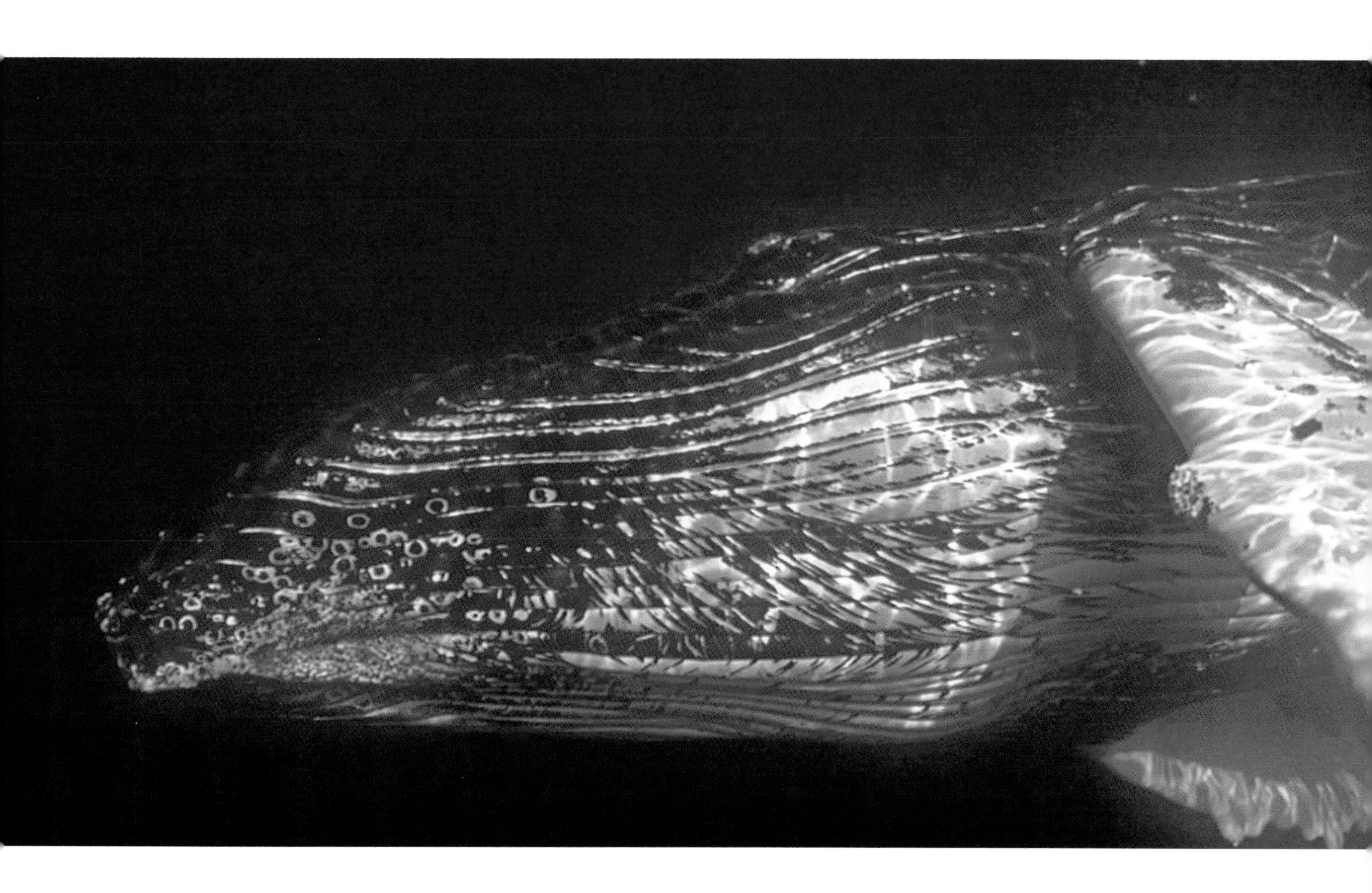

Magical Whale swam alongside me so that I was considerably closer than my own body length to him. With no apparent motion of his body, he slowly reversed until his massive head was next to mine. He stared at me with large, brown eyes. Then he lowered his head and remained motionless within arm's reach underneath me as he listened to my heartbeat.

The whale must have been familiar with boats. I'm sure it knew that I was a creature who had abandoned the safety of the boat to be in the water with him. His behaviour couldn't be explained by simple curiosity – the animal wouldn't have spent so much time dancing around me, investigating and signalling to me the way it had.

I have little doubt that the whale had tried to communicate with me, much as a dog might roll onto its back to expose its belly in a submissive gesture. My inability to connect was entirely my own shortcoming. Like most of my species, I am so used to using words for communication that the primeval ability to commune interspecies without verbal language has atrophied beyond my capabilities.

To dull my senses and soothe my mind, I concluded this whale was 'friendly', a whale that makes a habit of approaching boats in its feeding grounds. It had to be as simple as that.

Studying the details of the worn trailing edge of its fluke, and the massive scar on its lower right jaw, I guessed this whale was an older specimen. But we have no idea how old humpbacks can become because we've been killing the biggest and oldest ones for centuries.

I sent the photos of the unique patterns on the underside of the tail flukes to be identified by Allied Whale at College of the Atlantic in Bar Harbor, Maine, United States. They dedicated two students to the time-consuming task of matching his fluke to one of 6,000 tail flukes of North Atlantic humpback whales in their catalogue. Surprisingly there was no match. Not only was this not a 'friendly' whale, he had never been identified before, at least not by people that know to send their fluke photos to Allied Whale.

For months after this experience, Magical Whale consumed my consciousness and dominated my dreams. I studied the footage and became intimate with every part of Magical Whale's body. Just as that day on the beach with Elsa when the whale had lunged out of its world into ours, I had slipped overboard and submerged myself in the watery realm of the whales. Every night as I lay in bed I would think about Magical Whale as I had seen him, underwater. It was a soothing exercise. It calmed me to think about the whale. It gave me a sense of perspective on existence, on myself and on humanity as a whole. I had been touched in more ways than one by this whale's presence. Perhaps I had gone over to the other side. I was certainly no longer the same person I had been before my encounter with Magical Whale.

Had anyone else looked into the eye of a whale for such a prolonged period? Not a calf, or a whale stranded on a beach, or a harpooned whale, but an adult whale that voluntarily approached, thoughtfully to scrutinize a human eye to eye?

I trolled the internet, looking for footage that might be similar to my own. I bought all the whale films I could lay my hands on. It seemed that not only had I had a unique encounter, the mesmeric experience had been captured on my high-definition video camera underwater as well as the video camera on the boat. For a time I kept quiet about my experience with Magical Whale, finding it difficult to comprehend, let alone explain. For some months I didn't share the intimate moments revealed by that footage. I needed somehow to get a handle on what had happened. I relived that experience over and over and over again in my mind, trying to make sense of it, trying to understand the meaning of our meeting.

Something significant had happened, I just wasn't sure what.

It took quite a while after my encounter with Magical Whale to talk about the experience without becoming so emotional that I would cry. I knew, or felt, that I had been in the presence of an exceptionally intelligent being. At a loss to adequately describe how I felt to those closest to me, I said it was like looking into the eye of God.

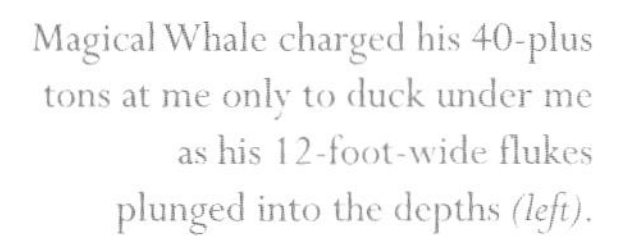

Magical Whale charged his 40-plus tons at me only to duck under me as his 12-foot-wide flukes plunged into the depths *(left)*.

Magical Whale danced beneath me for hours, but once, as he swam towards me on the surface, I crossed his path. We were on a collision course. There was no time for him to dive. He arched his back, extended his pectoral fins at right angles and rotated them so they acted as brakes while his massive fluke dragged through the water. I was less than a foot or so from being struck by a forty-ton leviathan.

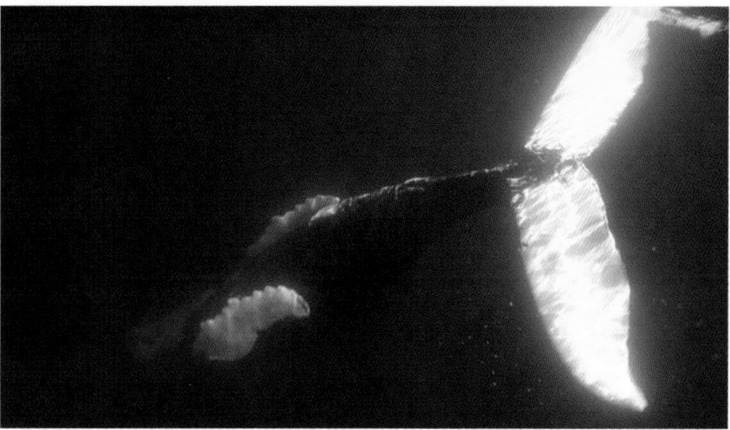

CHAPTER TWO

INTO UNCHARTED WATERS

My father, William Stevenson, was a foreign correspondent. Although I was born in Canada, I left when I was three and for most of my life I have been a nomad. Growing up in Africa, Asia, Europe and North America, I have more than one passport and can, or could, speak four languages fluently. Although I had vague aspirations to become an oceanographer or a game warden, I ended up studying postgraduate economics. Soon after graduating, followed by a short stint as an economist at the head office of a Canadian bank, I was recruited by the United Nations through the Canadian Foreign Service to work in Tanzania, East Africa. That same year my father wrote two bestselling books and my parents settled in Bermuda. After what seemed a couple of lifetimes later and an eclectic career that somehow combined my passions for adventure, flying, wildlife and mountains with making a living, I too moved to these islands, using Bermuda as a base to write about my travels. I met my wife, Annabel, a medical doctor from New Zealand, here in Bermuda. Already dating each other, I moved in with Annabel after shattering a vertebra in a motorbike accident. Two years later we spent our honeymoon in Bhutan and Upper Mustang.

My new-found interest in whales contrasted to my previous flirtation: the Himalayas. Bermuda had always seemed too small and unadventurous a place to remain permanently, but it was a perfect base, surrounded by family and friends, to write books based on my adventures around the world, including three books on the Nepalese Himals. For two decades prior to Elsa's birth, I had returned to those mountains almost annually. I was not interested in climbing to the summits. What fascinated me more was the life and culture of the people living in the valleys and mountainsides. My last extended trip to Nepal ended two months before Elsa was born.

The practicality of world travel diminished with the arrival of our first daughter. Although we did continue to make numerous journeys, as a necessity, they became less adventurous, and less exotic. Cast in the role of a stay-at-home father of a young daughter, my new role more than compensated for being grounded on a small island in the middle of the North Atlantic. In fact, Bermuda was a perfect environment for a toddler to grow up in. We could get outside all year around and there was our entire shoreline and inner reefs to explore.

While Annabel went to work at her medical practice each day, I remained at home with Elsa where I wrote magazine articles and put the finishing touches to books I had been working on. At the end of our drive is a long beach with a rock pool where Elsa and I searched for marine life. She began to bodysurf and discovered quickly that if the wave was too big for her little body, the safest and most prudent option was to duck under it. She learned to hold her breath and dive down before she could swim horizontally on the surface. It was on our beach and in the water that Elsa became fascinated with creatures, of any kind. Daily, we spent hours in the water or walking along the shore searching for critters.

The first whale Elsa and I saw close up was a dead humpback washed up on the rocks near our home. The baleen had been knocked out of its mouth and its body was distended by the expanding gases inside.

But, in researching the whales after Elsa asked me why the humpback jumped out of the water in 2006, I realized that where I live is, from a whale's perspective, the top of a mountain. Most of this mountain is now underwater, although Bermuda used to be several thousand feet above sea level, which would have made it a mountain bigger than Everest if you counted what lay beneath the surface of the ocean. Rising sea levels submerged this volcano and over millions of years layers of sand covered the seamount that became the Bermuda Islands. Then the wind, rain and waves wore the sandstone down again. Now the summit of Bermuda is barely 300 feet (90 m) above sea level. But, for a whale living in the middle of the ocean, Bermuda still looks like a giant mountain rising 14,000 feet (4,300 m) from the depths. And here in urbane Bermuda was a quest that surpassed anything I'd done before. I was living on a perfect platform providing a window into the mid-ocean lives of the humpback whales.

Soon after deciding to make a film about humpbacks, the first whale Elsa and I saw close enough to touch was, unfortunately, a dead one. It had washed up on our shores and had been towed out to sea after being identified as a dead adult female whale. Although discarded in the ocean, the carcass floated back to our shores the next day. There were no obvious injuries although on one side the baleen – the filtering structure of the mouth, sometimes called the whalebone even though it is made of keratin – had been knocked out of its jaw. By this time its pleated throat and stomach had become distended with the gases cooking inside and it was obvious by the protruding penis thrust outside its sheath that this was actually a male, not a female. I took Elsa out of school to come and witness this incident.

'Is it real?' Elsa asked, meaning, 'Is it alive?'

'It's real,' I replied. 'But it's dead.'

We studied the huge inanimate body solemnly for the longest time and the question kept coming back, 'Why is he dead?' No explanation I could give satisfactorily explained how this magnificent animal could end up lying on the rocks on his back, lifeless. Elsa was awestruck. Both of us were. The carcass bounced gently on the rocks as the waves nudged it backwards and forwards.

'It is it a boy or a girl?' Elsa asked.

'A boy,' I answered, knowing that response would mollify her.

'What now?' Elsa asked when she saw a boat approaching and divers tying a rope around its tail.

'They're going to tow him away, out to sea.'

'Why?'

'Because he's dead,' I answered.

She was silent. I could see she couldn't come to terms with the deadness of the whale. Lifeless on the rocks he wasn't real, like her plastic whales at home.

'Besides, the ocean is where he belongs,' I added.

When the boat dragged the dead humpback whale backwards and upside down beyond the line of reefs, Elsa asked quietly, 'Is he real now?'

'Yes, he's real now,' I replied.

It wasn't an auspicious start to making a film about humpbacks.

The camera gear I bought to start filming arrived soon afterwards, in February 2007. With Gates underwater camera housing, it seemed indestructible. It was heavy but compared to the size and weight of underwater filming cameras a few years ago, I had no right to complain. The underwater housing had two bulky tubes on top that extended forward of the housing. These were buoyancy tanks to offset the disproportionately heavy dome-port, or wide-angle lens, mounted on the front. I tried the camera out in a pool to make sure it did not leak. Now I was ready to try it in the ocean.

On the next calm day, I phoned a friend from what served as Elsa's bedroom at night and my office during the day. 'Nick, it's Andrew. I've done the weights and balance on the underwater camera housing in a pool – now I need to try it out in the ocean. It's a beautiful day. Can you come out with me on the boat...'

I stopped in mid-sentence and stood up as I gazed through the window over my computer screen at the ocean off Grape Bay. I saw a splash on the reefs just off the beach. 'Oh, my...!' I exclaimed, 'I don't believe it!'

'What is it?' Nick asked, alarmed at the tone of my voice.

'There are... humpback whales on the breakers!' There were three whales lying on their backs waving their pectoral fins, slapping them with vigour on the surface of the water. This was the exact spot where Elsa and I had seen the humpback breach almost a year earlier. It was an unbelievable coincidence. 'Nick, get over here quick!'

When Nick arrived the three whales were still lying off the breakers, repeatedly whacking the water with their 16-foot (5 m) pectoral fins as if they were beckoning me to join them.

We drove over to Hamilton Harbour to borrow a friend's boat, filled it with fuel and then motored all the way past Somerset on the west side of the Bermuda islands, and back east along South Shore to Grape Bay Beach, arriving there some hours later. By then the whales had disappeared. We trolled further down South Shore to St David's without seeing them. When we retraced our tracks and were cruising off Grape Bay Beach again we found the three whales. They must have been there all along. Nick turned off the engine. Heart pounding, I put on my wetsuit as the whales headed for the drifting boat and slipped nervously into the water just as the whales passed by. My heart was in my throat. Within seconds it was all over.

Things were looking good: three whales lying on their backs summoning me to get in the water with them, just when I had decided to try the camera out in the ocean. If that wasn't a good omen, I didn't know what was. Getting in the water with whales, camera rolling, first time out – this wasn't going to be as difficult as I thought.

Back at home I eagerly downloaded the footage on to my new Apple computer and checked the high-definition footage. The whales had passed by about 35 feet (11 m) away but on the computer screen they were barely discernible, vague cigar-shapes cruising by in the distance. With a 120-degree wide-angle lens, a 45-foot (14 m) whale 35 feet away looked more like a guppy.

It wasn't going to be as easy I had thought.

So that I would not lose whales spotted from the shore and to reduce the time it took to get into whale territory, Dr David Saul, ex-premier of Bermuda, lent me the use of his 25-foot (7.5 m) centre-console single-engine outboard boat, *Pheidippides*, and the use of his mooring in tiny Devonshire Bay. *Pheidippides* was small enough to get in and out of Devonshire Bay, in the centre of the South Shore, and big enough to patrol up and down the coastline into deep water. It had a tuna tower – a hardtop over the lower drive area with a control box for the engine and for steering. The tuna tower gave an extra bit of height over sea level, increasing the area of ocean visible to the eye.

I had zero open-water boat experience in Bermuda. David helped me to bring the boat from Hamilton Harbour along the South Shore to Devonshire Bay. He showed me how to navigate through the narrow gap in the first line of breakers where the surging swells pummelled the reefs a hundred yards offshore. Then we threaded the needle of coral heads for the remaining hundred yards to the entrance of the cove. At the entry to Devonshire Bay, there are sharp rocks that rise up above the surface at low tide. It takes deft zigzagging to avoid these razor-sharp pinnacles while piloting the boat into or out of the bay. The clearance under the hull of the boat at the entrance to the natural harbour at anything but the highest tide is a matter of inches. Entering or exiting the tiny cove necessitates raising the outboard engine as far as possible, while still getting some thrust out of the partially submerged propeller.

There are half-a-dozen small commercial fishing boats in Devonshire Bay. These are run by weathered fishermen who take their boats out of the bay to tie up to moorings they maintain in the deeper water off the South Shore. Tying their boats to their mooring lines just offshore, they angle for the larger predatory fish that are part of the food web surviving on the up-wellings of currents and tides that rise off the ocean depths and on to the Bermuda platform a mile or so offshore. The local fishermen, gutting their fish on a wooden table on the rocks at the edge of the bay, watched curiously as I unloaded the heavy camera equipment and waterproof containers out of my car to ferry in a small punt out to the boat.

Pheidippides had no working navigation equipment on board and the radio was suspect. Most of the time I was atop the tuna tower where I had no access to the boat radio below anyway. I carried my own tiny handheld GPS, a handheld radio and a cell phone.

I learned quickly. I had to. I began recruiting volunteers, even less experienced than I was, if that were possible. I was soon captaining the single-engine 25-foot (7.5 m) boat in the open ocean, looking for whales. Often I only had one other person with me. Sometimes I had my sister Jackie on board – she lived right on Devonshire Bay and it was easy to drop in to conscript her at the last minute.

One day she laughed out loud as we bumped into each other on the tuna tower while searching for whales: 'This reminds me of when you ran your safari business in Tanzania and we'd be in your Land Rover looking for elephants. Nothing's changed, except instead of finding elephants in a Land Rover we're in a boat looking for whales.' I had quit the United Nations after two years and started a safari business in the Selous Game Reserve in southern Tanzania. It was the largest reserve in Africa and a real wilderness where every morning before dawn I'd take clients out for a walking safari. Jackie had a point. I'd come full circle. Fast forward a couple of decades and instead of walking safaris in the East African bush, I was swimming with whales in the open ocean. Or at least, I was trying to.

Having *Pheidippides* moored in Devonshire Bay meant that I was in whale territory within a couple of hundred yards and a couple of minutes from the mooring. There were also distinct disadvantages. We had to row out to the boat in a tiny punt, transporting crew and heavy equipment. Refuelling meant ferrying fuel to the boat in containers. The closest marine gas station in Somerset was a half-day return journey and would empty a quarter of the tank. There were many days in that first season when the fishermen didn't go out because of the waves breaking through the reefs into our protected little bay. The fishermen may have thought I was a fool to risk going out on those days, but they thought I was foolish anyway for trying to swim with whales in the open ocean. They were intimately acquainted with what lurked beneath the surface: sharks, in particular the tiger shark. From the fishermen's point of view, swimming in the same waters with sharks or whales didn't make any sense at all.

I was scared too, but not for the same reasons. There were many nights when I couldn't sleep, worrying about taking Pheidippides out of the bay the following morning in conditions that were often marginal. The technique to motor out of Devonshire Bay and beyond the reefs required determined aggressiveness and a steady hand. The waves that came through the outer reefs picked up speed and height as they left the deeper water off the breakers and entered the shallows. The bow of the boat had to meet the waves head on to break through these heaving swells, like a lifeboat cutting through the surf. Not enough speed – or hitting the waves at an angle – would turn the boat sideways. There wasn't enough room to manoeuvre the bow round again and the next wave would inevitably strike the boat broadside, sending it on to the rocks on either side of the entrance.

David Saul was unperturbed about the risks presented to his boat. 'Don't worry about *Pheidippides*,' he told me more than once. 'Just make sure you swim to shore safely. I've had that boat twenty-five years and it doesn't owe me anything.'

Often it was difficult to tell where the breakers were and where the gap through them lay. Lining up the pillars outside the front door to David's house, the aligned columns acted as a perfect guide to getting through the outer breakers. I suppose there was a certain symbolism in that.

There were days out on the open ocean when the tuna tower lurched at such an angle that I was sure the boat was going to capsize. At the end of the day, when we finally reached our mooring within the protected waters of our cove, we were often drenched and exhausted. Sometimes the small cabin at the

Studying whales or dolphins from above the surface of the ocean is akin to studying elephants by examining the tips of their trunks when they come to a waterhole to drink. Spending so much time in the water, I had a window into the whales' pelagic migratory behaviour.

We also occasionally saw Cuvier's Beaked whales *(lower image above)*.

bow of the boat was full of water. On one outing the boat's heavy ploughing into the waves had knocked the two exterior bow headlights into the cabin and water had poured in through the two portholes.

At times the weather was calm but heavy swells from a far-off storm battered our shores. These were the worst waves – as they crossed the shallow waters inside the outer breakers they rolled in rapidly and in close succession. Even in calm weather, if there was a distant storm, we were confronted by dangerous surf that bore down on us relentlessly.

And then one day the engine failed as I slowed down outside the breakers to line up on David's pillars. Just as I positioned the boat between the reefs, the engine cut and we were left bobbing up and down helplessly as the currents tossed us about tantalizingly close to the breakers. It took some minutes to start the engine again but even that delay could have been catastrophic. We were fortunate it was calm with no wind or waves to push us onto the coral heads, which would have demolished the hull very quickly. If that had happened, we would have been lucky to have had the chance to follow David's advice and swim for shore, leaving his boat a wreck, the latest victim of Bermuda's notorious reefs.

Once, coming back at dusk in poor light, I miscalculated the position of the reefs in the shallows and hit one with the propeller. Fortunately, we still had enough thrust to get back to our moorings. That required a US$1,200 replacement of the stainless-steel propeller.

There were times in my life, especially in my twenties, when I thought I was immortal and I took unnecessary life-threatening risks. I can count the cat's nine lives I've used up. But now, having a small child at home, I was acutely aware of the gambles I took. There were plenty of possibilities for tragedy swimming for extended periods 15 miles (24 km) offshore and I did whatever I could to reduce the chances of losing my life. The one thing that I worried about the least was the risk posed to me by the whales.

The only way to get close enough to a whale to film it is to let the whale swim to you. There is no point in chasing a whale in a boat and even less point chasing it in the water. I cannot swim faster than a whale and so it becomes a numbers game of getting in the water whenever a whale looks like it will come to investigate. It takes perseverance and patience, as well as resilience to the effects of cold water. The technique I used to get in the water with a whale was to follow it at a discreet distance in the boat, watch its behaviour for an extended period and then hope that it might eventually become curious enough to turn around and investigate. If it seemed the whale was heading towards us, I'd slip into the water. I was probably bobbing around in the water unsuccessfully at least fifty times for every good encounter I experienced.

To be less intimidating to the whales we tried to have the engine off when I was in the water waiting for the whales to approach. Inevitably the boat would float away from me, pushed by the wind, waves and currents. More than anything, I worried that the boat engine might not start again while I was still in the water some distance away. Offshore in the open ocean and in currents that were often more than 4 knots, there was little chance of swimming to an incapacitated boat being blown by the wind in the opposite direction and drifting away from me. There was even less chance of swimming 15 miles back to Bermuda unless the tides and currents aligned in just the right direction. I experimented with hanging on to a rope tied to the boat, but decided that was even more dangerous if a whale swam between me and the boat.

Although the first occasions in the water in the presence of whales were scary, that fear wore off as I realized that they posed no threat to me. More threatening are the Portuguese man-of-war jellyfish, floating on the surface of the ocean with their lethal trailing tentacles dangling below. Sharks were a potential threat although I didn't consider them crazed man-eating machines that would suddenly appear out of the depths to chomp me. On numerous occasions we saw different species of shark as we cruised around looking for whales but none came to investigate or bother me.

Bermuda is at the same latitude as Charleston, South Carolina, and although the Gulf Stream keeps our waters relatively warm, they aren't Caribbean water temperatures, at least not at the end of winter when I was in the water looking

North Atlantic Bottlenose dolphins are often seen in the canyon between the Bermuda platform and Challenger Bank (left).

(right page) The White-tailed Tropicbirds, known locally as longtails, appear around Bermuda at the same time as the migrating whales.

The first day I was ready to use my underwater camera there were three humpbacks off our beach at Grape Bay, lying on their backs and whacking the water with their pectoral fins as if beckoning me to get into the water with them.

Returning from a long day looking for whales I'd find three-year old Elsa standing in the surf on Grape Bay Beach, waiting for me to come home. There isn't a lot of room for error getting in and out of Devonshire Bay with sharp rocks and coral heads protecting the entrance. The local fishermen told me I was crazy to go out in the some of the conditions we set off in.

for friendly whales. The water temperature is around 65°F (18°C) and the air temperature about the same. Long sessions in this water, climbing into and out of the boat with a heavy camera, and standing around in a soaked wetsuit took a toll on my muscles. Most of the time I wore a wetsuit in the water, but that depended on how much time I had to get in. There was no point putting on a wetsuit on the boat because my body overheated in the sunlight and humidity, and I became dehydrated. If a whale surfaced unexpectedly nearby and seemed to be heading towards the boat I simply stripped out of my clothes and slipped into the waters in my underpants. The waters in Bermuda in March and April are cold enough that floating in the water without a wetsuit becomes dangerously debilitating in a relatively short period of time. There was always the risk of cramp and I discovered that using my fins to leverage the heavy camera left or right had the effect of cramping my hamstrings.

I was never in the water with scuba tanks. There were two reasons for this. First, humpbacks blow bubbles at each other as a threat display. I certainly didn't want to threaten them. The second reason was safety. While floating on the surface and free-diving for short periods at a time, the crew on the boat could keep track of where I was. If I was submerged for any length of time with tanks, it would be impossible for the boat to be aware of my position. And I had noticed something about my own behaviour: once I had a whale anywhere near me I was totally focused on filming the whale and was oblivious to anything else.

I elected to go alone in the water because I didn't want anyone in the viewfinder and also because, curiously enough, I felt safer. With only one diver in the water, the whale always knew where I was. Besides, despite their initial enthusiasm, no one else seemed to want to get into the chilly waters time after time in the vague and mostly vain hope that a whale would come by.

In such circumstances I was pushing the limits of my luck. I was, after all, swimming in the open ocean, often 15 miles or more from Bermuda. I had already used up my allocated nine lives, most of them in Africa. And now I had a wife and daughter to worry about.

The first season we patrolled Bermuda's South Shore at least a mile or two from the breakers in waters that ranged between less than a hundred to several hundred feet deep. On average, one out of every three occasions that we went out looking for whales along the South Shore we found them during that first year. Most often they were moving west to east, which made sense. That was, after all, the direction of their migration. Sometimes it was a mother and calf, sometimes a threesome. Sometimes it seemed that a mother and calf might be off the same point of land for more than one day. If we drove the boat towards the western edge of the island and further offshore, it seemed we encountered more whales. Over time, as I gained confidence, we started meandering further and further offshore. Without a working marine map and GPS depicting the ocean floor's depths beyond the couple of hundred feet that the depth sounder sometimes registered, we were searching blind, and our only term of reference was land.

Local fishermen in Devonshire Bay had told me that I would find plenty of whales at full moon when the krill would come up to feed on the phytoplankton. On a calm morning the day of the full moon in late March, I left Devonshire Bay on *Pheidippides* with one of my regular volunteers, Camilla Stringer.

Visibility was perfect outside the reefs, the water so clear that it seemed we were floating on air some 70 feet (21 m) above the coral. We spotted several spouts on the horizon, all heading west. The water was so calm we were able to see a whale's 'footprint' as it swam at about 5 knots just below the surface. It was on a mission, heading in a straight line somewhere. We paralleled the whale 30 yards (27 m) to its side, and eventually it came alongside and rode the bow wave of the boat. We travelled together along South Shore until we came to Sally Tuckers off Somerset. The water was calm as an oil slick up to that point and yet ahead of us was a ruffled ripple line extending some hundreds of yards along the 'edge' of Sally Tuckers. It was clearly an up-welling of currents. On top of the ripple line was a streak of foam. We threw a funnel net with a bottle attachment overboard and collected a water sample before continuing north to the other side of the ripple line. While the water on one

One day we threw a funnel net with a bottle attachment overboard and collected a water sample. The next day, under the microscope, we determined that the Petri dish was full of decapods, copepods, tiny fish and fish eggs.

One day we found a *Histioteuthis*, a deep-sea squid. The mantle, or body, had been bitten off, making it look octopus-like. This was only the second time a *Histioteuthis* had been found in Bermuda *(bottom)*.

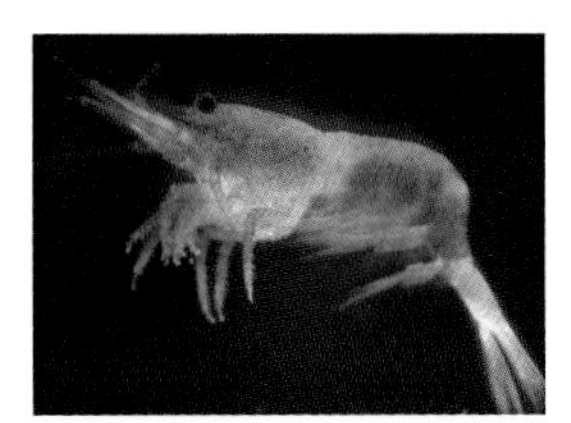

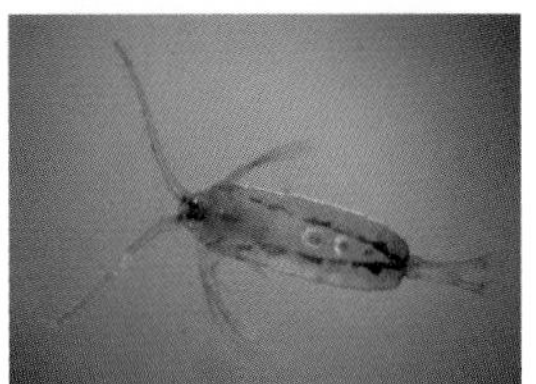

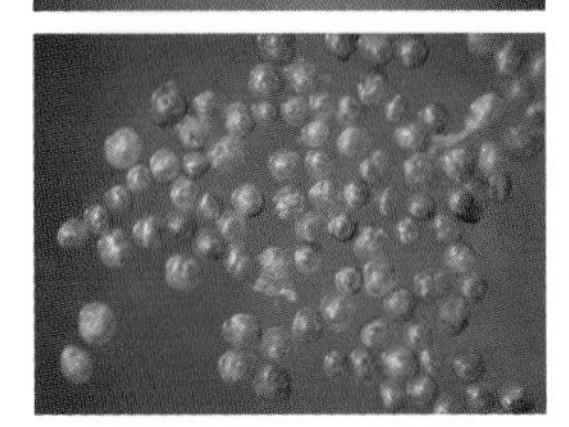

side had been crystal clear, on the other it was murky. Over the next hours we saw scores of whales, apparently feeding on the murky side of the up-welling. I tried getting into the water to film these whales but the water was so gloomy that I couldn't see them despite one apparently swimming right under me. That afternoon it seemed whichever way we turned the boat we saw whales, and always humpbacks. Sometimes they appeared out of the blue right beside us. I guessed we saw twenty, perhaps as many as thirty, whales that day.

The next morning, with another of my dedicated volunteers, Judie Clee, we took the bottled sample of water to the Bermuda Institute of Ocean Sciences where marine scientists emptied the contents into a large Petri dish. Right away we could see with our naked eyes what appeared to be small shrimp-like crustaceans. Under the microscope we determined that the Petri dish was full of krill, decapods, copepods, tiny fish and fish eggs. The scientists were even more surprised when I informed them I was only a few miles offshore from Bermuda and that I had towed the funnel net for less than half a minute. The whales had clearly been feeding on a nutrient-rich environment fuelled by an up-welling of deep water off the seamount.

Dr Clyde F. E. Roper, Emeritus Zoologist at the Smithsonian Institution commented on my description of events:

Your observations with regard to the full moon are interesting. In clear oceanic waters, as demonstrated in the research Bob Gibbs, others, and I did in the 70s off Bermuda, a full moon depresses the mesopelagic vertical migrators (macro plankton and nektonic fishes, squids, crustaceans) because they adhere to a specific ambient light intensity, not depth, when they migrate from mid-depths toward the surface. Thus, on the full of the moon, they do not come up so far as they do in the dark of the moon. That dynamic obviously changes in the case of strong up-wellings in highly enriched, productive waters where the primary producers and micro plankton don't give a hoot about light intensity differences or depth. They just keep on producing, and the consumers concentrate where the food is. In fact, if the biomass were concentrated heavily enough in the upper

100–150 m, that density could reduce the amount of full moon light that penetrates, and, consequently, those predators that normally would stay deeper can now enter the smörgåsbord without sunglasses!

One day we found a *Histioteuthis*, a deep-sea squid. The mantle, or body, had been bitten off, making it look octopus-like. This was only the second time a *Histioteuthis* had been found in Bermuda. Being the favourite food of some odontocetes, or toothed whales, which tend to dive deeper and are known to be especially fond of squids for a meal, this squid fitted in perfectly with our sighting the day before of a Cuvier's beaked whale.

While conventional wisdom indicated that the humpbacks migrated by Bermuda from late March through April, in that first year we encountered humpbacks from February well into May. I also had anecdotal evidence that whales were around Bermuda much earlier in the winter.

I left notices, tacked on to trees and fences and in mailboxes, all over South Shore, asking people if they spotted whales to phone me on a dedicated cell phone number: 77-SPOUT. Having hundreds of pairs of eyes onshore looking for whales is a lot more efficient than a couple of pairs of eyes bobbing on top of the tuna tower on a boat. It was important to document all the whale sightings, especially outside of the regular migratory period, so that we could better understand what they were doing around Bermuda.

On a couple of occasions we followed a group of whales for several miles moving in the direction of their travel along the south-west to north-east alignment of our South Shore. One memorable time we followed seven whales heading from Challenger Bank past the point at Sally Tuckers and as they continued towards the south-west breaker and along South Shore. Eventually we left them at dusk and returned home in poor light through the reefs back to Devonshire Bay.

Early the next morning I had a call from a spotter on the hillside overlooking the waters off Devonshire Bay. He reported several whales just offshore. Predicting that the whales had spent the night in the protective waters off Devonshire Bay before continuing their journey north, we left Devonshire Bay, turned left and headed for St David's to our east. We didn't find the whales.

My initial predictions of having an easy time filming whales were sadly way off the mark. Despite the good omen those three whales seemed to represent on the first day I tried my camera in the ocean, I didn't manage to get any closer to whales that season. And I never did figure out what those three whales beside the breakers were doing, whacking their pectoral fins for a couple of hours while lying on their backs. Besides opportunistically feeding here in Bermuda, I had no idea what the humpbacks were doing in the area beyond migrating on their way north. It seemed the more time I spent on the water observing the humpbacks, the more questions I had.

At the end of April I was physically exhausted. I was also emotionally spent from the expectations I placed on myself to film whales underwater and make a thirty-minute documentary.

The season was rapidly turning into a bust.

And then I had my encounter with Magical Whale.

CHAPTER 3

IDENTIFYING CANDLE – FLUKE IDS

After my encounter with Magical Whale and once our 'whaling' season was over in May 2007, I was able to organize all my still photographs of the whales' flukes. Having sorted through the images, I realized I had photographed the other whale that had intruded on my intimate moments with Magical Whale several times over a three-day period that week. We nicknamed him Candle because of the black candle-like marking on his all-white flukes.

Armed with my best photograph of the underside of Candle's tail flukes along with fourteen other fluke identifications (IDs) I made that season, I flew to Bar Harbor, Maine, to meet Dr Steven Katona of College of the Atlantic, and Rosemary Seton and Judy Allen of the marine mammal research group, Allied Whale.

Researchers at Allied Whale have discovered that individual humpbacks can be identified by the unique black-and-white patterns on the underside of the whale's fluke, similar to fingerprints for humans. Photographs of the underside of a humpback's tail are called 'fluke IDs'. Allied Whale's North Atlantic Humpback Whale Fluke Identification Catalogue has recently increased rapidly in size – it has now identified over 7,000 individual whales from the North Atlantic alone. This catalogue includes 146 whale fluke photo IDs made here in Bermuda from 1968 to 2006. As mentioned, students and staff had tried to match Magical Whale's flukes to those in the catalogue but had come up with nothing. I was surprised. I thought Magical Whale was a 'friendly', a whale that made a habit of 'mugging' boats by coming up to them. But, it seemed that Magical Whale had never been photographed before, and the mystery of who Magical Whale was and what he was doing with me deepened.

After a day spent with members of the faculty and students, I settled down in the Allied Whale office alone that evening to go through the North Atlantic Humpback Whale Catalogue (NAHWC) to see if I could identify Candle, the whale that had approached me when I was with Magical Whale. In Bermuda, I had assigned Magical Whale our catalogue No. 0001, and Candle was No. 0002. Judy and Rosemary left for home asking me to turn off the lights and lock the offices when I left for my motel.

I started with the oldest photos at the beginning of the catalogue in the 'Type One' category (see Appendix). These were whales with less than 25 per cent black markings on a white background. By chance, within minutes of opening the fluke ID catalogue containing thousands of photographs, I had made the match and identified Candle.

'Are you sure?' Judy Allen asked when I phoned to tell her. She immediately drove back to the office. There was no doubt at all, a perfect match. Candle had been photographed in 1978, twenty-nine years earlier, by Dr Hal Whitehead in Newfoundland and had never been seen again until I took his photograph on 24 April 2007. Presumably this whale had migrated southward and northward for upwards of thirty years and had never been photographed in the interim. Matching Candle so quickly was a small consolation for not matching Magical Whale.

Armed with their encouraging feedback on my observations and with the scientific papers Allied Whale had given me, I returned to Bermuda. I was a blank slate, a receptacle for any and all ideas. I had no training and therefore no dogma. But after reading all the scientific papers I could get my hands on, besides vague references to these islands being a navigational waypoint, it seemed no one knew what the whales were doing migrating right by Bermuda. If turtles and birds could navigate thousands of miles with a brain a tiny fraction the size of a humpback's, then surely the humpbacks didn't need a physical reference point like an island in the middle of the ocean just to get to where they were going. There had to be more to it than that.

By now I was thoroughly hooked on humpbacks. Getting a glimpse through the unique window that Bermuda provided into their social behaviour during their mid-ocean migratory crossings was fine, but I needed to be more familiar with their annual lifecycle elsewhere to really know them. To understand them better I needed to travel to their feeding grounds.

North Atlantic humpback whales spend spring, summer and autumn in high latitude feeding areas, dispersed on the Gulf of Maine, Bay of Fundy, Gulf of St Lawrence, Newfoundland/Labrador, Greenland, Iceland and Norwegian waters, foraging for food to sustain their winter migration to southern waters where there is little or no sustenance. By the end of the summer they are fat and happy from feeding in the fecund waters up north.

Although it can vary from year to year, many humpbacks clearly maintain a high degree of site fidelity to their feeding grounds. If their food is scarce, they may move further along until they find food. Over the years many of the whale-watching tour operators in the humpbacks' feeding grounds recognize the same whales returning to the same bays and harbours. Because the whales tend to return to the same coastline, research groups and naturalists as well as whale-watching operators are able to document the mothers with their new calves, and slowly build up a history of the whales' familial ties.

Halifax, Nova Scotia, was an easy and direct two-hour flight north from Bermuda. It had the added benefit of being in the same time-zone. Annabel, Elsa and I followed the southern Nova Scotian coastline west and then cut north across to Digby Neck, a long spit of land that extended to the very entrance of the Bay of Fundy. Two short ferry rides later and we were on Brier Island, the childhood home of Joshua Slocum who 1895–8 became the first person to circumnavigate the world alone. On Brier Island we didn't have to look for the whales – they seemed to be looking for us just outside the small harbour of Westport. They were so friendly there that I wasn't sure if we were the ones watching the whales so much as the whales were studying us. We were 'mugged' often by fat and happy humpbacks determined to interact with the people on the boat. On a number of occasions they approached the lobster fishing boat-turned whale-watching boat to spy-hop (bring their head vertically out of the water) or roll around beside the boat, with an eye or both eyes out of the water, looking curiously at us.

I have a photo of a whale waving his 16-foot (5m) pectoral fins across the gunwales and stern of our boat and making contact with the naturalist on board who had extended her hand over the side. The brief touch made her emotional to the point of tears, despite the fact that the palm of her hand was shredded and bleeding profusely from the razor-sharp barnacles on the leading edge of the pectoral fin.

It was clear to me why marine biologists knew so much about the whales in their feeding grounds. It was easy. Even with Bermuda as a platform to gain an insight into their mid-ocean lives, it wasn't anything like the experiences we had at Brier Island. If I could have the same close encounters with whales on such a regular basis in Bermuda I would have a much better idea of what they were doing there.

(previous page) The humpback deliberately rolled onto its back and waved its pectoral fin alongside our boat, giving us a high-five.

Elsa using her fingers as imaginary dinosaurs, even as the humpbacks approached us *(left)*.

I identified Candle (above) within minutes because I started at the beginning of the catalogue, with one of the oldest flukes.

We rented a camper-pickup and camped on the coastline of Nova Scotia to look for Magical Whale.

Two short ferry rides later and we were on Brier Island, the childhood home of Joshua Slocum, who became the first person to circumnavigate the world alone. We didn't have to look for the whales -- they seemed to be looking for us just outside the small harbour of Westport.

Although this can vary from year to year, the humpbacks maintain a high degree of site fidelity to their feeding grounds. Over the years many of the whale-watching tour operators in the humpbacks' feeding grounds recognize the same whales returning to the same bays and harbours.

After Brier Island we rented a camper-pickup and camped on the coastline of Cape Breton. We saw many pilot whales, but no humpbacks. At the end of our summer holidays we flew back from Halifax directly south to Bermuda. As winter approaches, most but not all of the North Atlantic humpbacks leave their feeding grounds and migrate southward to the mating and calving grounds of the Antilles in the Caribbean. Their exact migration route south is unknown.

I would have to wait until next spring to see the humpbacks again because they do not return via Bermuda in the autumn. That fact alone gave us a clue as to why they head close to us on the northward leg. One obvious difference was the fact that they were hungry going north and they weren't hungry going south. They had little incentive to scrounge for food on a mid-ocean seamount on the southward journey when they were stuffed to the gills, or more aptly, had put on all the blubber they needed. Perhaps that's all there was to it after all – they came by us to feed. Heading northwards from the Caribbean, Bermuda is the first in the chain of New England Seamounts that the whales would encounter in the Sargasso Sea. These seamounts are productive areas because they influence ocean currents, creating up-welling with retention zones and eddies. The Sargasso Sea was thought to be relatively unproductive, but a series of food 'oases' created by seamounts could provide important stepping stones for any hungry animal passing through.

Whether the time spent opportunistically feeding on the less productive seamounts for appetizers rather than moving to more productive feeding grounds up north was worth the delay was a good question. It didn't seem to me that it was worthwhile for the humpbacks to be stopping just for occasional appetizers in the up-wellings on a mid-ocean seamount versus continuing their migration as quickly as possible for the main entrée in their feeding grounds.

Was there another reason why they came by Bermuda in the spring?

CHAPTER FOUR

IN THE BREEDING GROUNDS

To find out more about the humpbacks, I flew down to the Caribbean in early 2008 and joined a commercial tour operator on a 'swim with the whales' excursion. Humpback whales are protected all over the world and, as far as I know, there are only two places in the world where legal, government-sanctioned, swim-with-the-whales tours are offered by commercial operators: Tonga and the Dominican Republic. All I had had to do was get there and pay some thousands of dollars to have experts get me and some ten to twenty additional paying clients into the water with the humpbacks. Two or three commercial tour operators have been 'working the whales' here for as long as two decades.

Located 60 miles (100 km) north of the Dominican Republic and approximately the same distance from the Turks and Caicos Islands, Silver Bank is one of numerous breeding and calving zones of humpback whales. In 1996 the Silver Bank Sanctuary was enlarged and renamed the Sanctuary for the Marine Mammals of the Dominican Republic. The jurisdiction of the sanctuary now encompasses Samana Bay and the northern and eastern coastlines of the Dominican Republic, a frequently travelled area of the North Atlantic humpback whale while en route to various breeding and calving locations in the Antilles. These breeding and calving grounds include the waters of the Dominican Republic (primarily Silver Bank, Navidad Bank and Samana Bay), Mona Passage (Puerto Rico), Virgin Bank and Anguilla Bank.

It's an overnight trip on a live-aboard dive-boat from Puerto Plata on the north coast of the Dominican Republic to Silver Bank. The boat arrives at its destination early the next morning so the captain can navigate through the reefs. Near our mooring was a protective ring of reefs some hundreds of yards to one side of the boat. The only landmark in a featureless horizon of ocean was the rusting hull of a listing shipwreck on the ring of reefs. As far as the eye could see there were whales, humpbacks and nothing else. During a winter season, 5,000 or more humpbacks pass through the 20 square miles (52 square km) encompassing Silver Bank, the largest concentration of humpbacks in the North Atlantic Ocean, if not the world.

I had misgivings about these tours – that they might be harmful to the whales at worst, or disturb them at best. Being a tourist on a 'swim-with-the-whales' holiday excursion didn't appeal to me but, not being a marine scientist, it was the only way I would be able to get an underwater insight into the humpbacks' lives in their breeding grounds. On Silver Bank the two or three operators, with up to two dozen clients, swim with thousands of whales. The tour operators' activities are restricted and monitored. Being 80 miles (128 km) out to sea prevents the casual inexperienced tourist boat from coming over from the mainland. There are only three licensed tour operators and they brief the tourists on how to 'swim-with-the-whales', or 'float-like-a-jellyfish', which is a better description. From what I have observed, there was no obvious harassment of the whales.

Located 100 km north of the Dominican Republic and approximately the same distance from the Turks and Caicos Islands, Silver Bank is one of numerous humpback breeding and calving areas. This wreck is the only visible landmark on an otherwise empty horizon, indicating an area of shallow water where mothers and calves stay for protection.

A calf opens its mouth wide to swallow water, filling its pleated throat.

It is easy to have close-up, in-the-water encounters with the humpbacks on the Silver Bank, especially the curious calves. This calf is less than two weeks old and not much more than 12 feet long. It must come to the surface to breathe every three or four minutes before diving back down to its mother.

On the positive side, swimming with whales is likely to make the most hard-nosed participant an active whale advocate. It also provides a poor country with a steady source of foreign exchange, placing a value on the humpbacks alive rather than hunting whales as the islanders of St Vincent and Grenadines do. On the negative side, the long-term effect of habituating whales, especially calves, to boats and human activity could be problematic, especially when they move on to their northern feeding areas with increased potential for collision with boats and lethal propeller cuts as well as entanglement in ropes and buoys.

My knowledge and first-hand experiences increased exponentially during the week I was on Silver Bank. Having spent 300 hours on the water and many sleepless nights in Bermuda worrying about being the responsible captain of a small boat in the middle of the ocean, being on Silver Bank was magical. I had no responsibilities and every day over the five days we were on Silver Bank I observed aspects of humpback behaviour I had not seen before.

Why the humpbacks migrate to warmer waters in the winter is still a matter of speculation and debate. Some marine scientists say it is because the calves must be born and weaned in shallow, protected and clear waters where there are no sharks or orcas (also known as killer whales, although in fact orcas are dolphins, not whales). Some say the warm water is necessary for the calves because they are born without a sufficiently thick insulating layer of blubber and that putting on weight is easier in warm waters. I've heard others say the humpbacks migrate south to get rid of the barnacles which fall off them in the warmer waters. Maybe it's a combination of all or some of these reasons.

During their winter gathering in the Antilles, the humpback males become vocal, active on the surface and highly competitive. In anthropomorphic terms, after spending the summer working hard to earn the cash, they are now in holiday mode and looking to splash out, burning up their hard-earned blubber, essentially to procreate. Like young men bar-hopping on a Saturday night and looking for a date, the male humpbacks are looking for a chance to mate. The females are more selective, waiting for the biggest, most powerful male to mate with.

Humpback whale experts seem to agree that the mating ritual starts with dominant male humpbacks securing their position next to a female and protecting that place from any challenging whales in the area. Some of these females may be with a newborn calf. If another whale challenges a male escort with a female, the escort will display postures to warn off the competition. If a challenger is unsuccessful on its own, it may join a competitive group to use their greater numbers in an attempt to dislodge the dominant escort. These competitive groups are obvious on the surface as they breach or whack their pectoral fins to get the attention of the female and intimidate the escort. The escort in turn swats at each approaching male, rams into them, sideswipes them with its tail, and does anything to keep its coveted spot next to the female. This can become a startling, and sometimes even deadly, spectacle when a dozen or more 40–50 ton challengers jockey for position near a female.

Being in these waters with the whales during their breeding season gave me new insights into their lives. Not only did we have remarkably close encounters with whales, we could also hear the humpbacks singing. There was a constant background noise of whales: groans, moans, squeaks, creaks, whoops and everything in between, and sometimes all together. Diving deeper was like turning up the volume on the speakers. A couple of times we located a singer. The singers were usually stationary, staying in one geographical location for extended periods of time, head down, about 50 to 60 feet (15 to 18 m) below the surface. If it hadn't been singing, I might have concluded the singer was sleeping. The singer maintains this position until it surfaces to breathe and then immediately resumes this position. Scientists have also discovered that singers can travel steadily while singing and may move tens, if not hundreds, of miles during a song session.

One evening, as the sun was setting, we came across two humpbacks head to head about 90 feet (27 m) down, just off the bottom. I was given permission to free-dive down to them with my camera. Each time I descended I could hear a chirruping sound that became louder as I approached. I dived to them a dozen times and on each occasion the very soft chirruping became clearer the closer

I got. I don't have any doubt that one or both of these whales were communicating to the other in very quiet tones. I wondered if they were about to mate. Unfortunately, as the whales were almost on the bottom, I did not swim under them to tell if they were male or female. As the sun set we left them to board our mother boat. Soon afterwards the two of them gave a synchronized display of upside-down lob-tailing.

Some weeks later, at the invitation of Flip Nicklin, the National Geographic photographer/filmmaker, I flew to the Whale Quest Conference in Maui, Hawaii, to learn from the experts about the Pacific humpbacks.

Every morning I woke up early in Flip's guesthouse to walk down to the beach at Kapalua where I immersed myself in the water. If the singing on Silver Bank was loud, it was nothing compared to the humpbacks singing here. With my head just below the surface I could easily hear their songs. Diving a few feet further and the songs became even more intense. It seemed there were hundreds of whales singing simultaneously.

Over the centuries sailors recorded that while lying in their bunks close to sea level they heard the ocean 'sing'. This is more likely to have been humpbacks singing. The first recognized recordings of humpback songs were made in the early 1950s in Bermuda by Frank Watlington, an electronics specialist responsible for recording and identifying underwater sounds using hydrophones (underwater microphones). In the 1950s and 1960s Watlington was stationed at the US Government's Naval Underwater Systems Center sonar listening base in Bermuda. His job was to monitor underwater sounds for the tell-tale signs of Soviet submarines. Each spring, unidentified sounds were recorded by the sonar devices, much to everyone's bewilderment. It was Watlington who solved the mystery, identifying the sounds as humpbacks. His recordings, some from as early as 1952, drew the attention of scientists including Dr Roger Payne, and set in motion a series of studies of humpbacks and how they communicate.

Roger Payne and Scott McVay first described humpback whale song in a journal article, 'Songs of the Humpback Whale', in Science in 1971. Among their findings was the fact that humpback whales produce a series of beautiful and varied sounds for a period of seven to thirty minutes and then repeat the same series with considerable precision. They called such performances 'singing' and each repeated series of sounds a 'song'. They found that all prolonged sound patterns of the humpbacks are in song form. They also discovered several song types around which whales construct their songs, but with individual variations around a very rough song pattern. These songs are repeated without any obvious pause between them so that song sessions may continue for several hours. The recordings of Frank Watlington and Dr Roger Payne were included as a freebie in *National Geographic* magazine and millions of readers were treated to 'Songs of the Humpback Whale', the 1979 sound sheet.

Through the 1970s and early 1980s, Roger and his wife Katherine described the structure and dynamics of the humpbacks' songs. In 1978, the Paynes began a study in Hawaii with Jim Darling as an assistant. This work with the Paynes eventually led to Jim's PhD work on the behaviour of humpback whales. At that time, rarely had a singer been located and identified and virtually nothing was known about the singers. I was fortunate that Dr Darling was working with Flip while I was in Maui.

Through decades of research, Jim has found out much more about the general structure of the song as well as the basic characteristics of singers. These characteristics, combined with limited observations of singing whales, have led to several ideas as to the function or role of the humpback whale song on the breeding grounds. Jim's research tests these ideas through intensive observation and measurement of the behaviour of singers – and their interactions with other whales. Recent advances in technology have enabled profound new discoveries in this field with the use of suction-cup tags, bio-acoustic tags, Crittercams and other devices.

Marine mammals, such as whales, dolphins and porpoises, are much more dependent on sound than land mammals; their vocalizations play an important role in everyday life functions including communicating, socializing, courtship and mating, territorial defence, care of young, foraging and environmental

One evening, as the sun was setting, we came across two humpbacks head to head about 90 feet down, just off the bottom. Each time I free-dived to them I could hear a quiet chirruping sound that became more obvious as I approached. Soon afterwards the two of them gave a synchronized display of upside-down lob-tailing. To date, no one has witnessed humpbacks mating, or giving birth.

Humpback whale experts seem to agree that the mating ritual starts with a dominant male humpback securing its position next to a female and protecting that place from any nearby challenging whales. Some of these females may be with a new-born calf *(right)*.

A humpback whale calf is between 10 and 15 feet (3-4.5 m) long at birth, smaller than the pectoral fin of its mother, and weighs up to 1 ton (907 kg). It feeds frequently on the mother's rich milk, which has a 45 per cent to 60 per cent fat content. Unlike the adults, the exuberant calves didn't always seem to be completely in control of their rapidly growing bodies.

sensing. The mechanisms used to produce sound vary from one family of cetaceans to another although, unlike us humans, cetaceans don't have vocal cords. Scientists think humpbacks have a larynx and although they are not entirely sure how they produce sounds, they may produce them just like us. The speed of sound in salt water is roughly four times the speed of sound in air, making it the most effective and efficient tool for communication in the ocean, and the only means of communication over any distance. Other senses are of limited range and effectiveness in the water. Sight is limited for marine mammals because of the way water absorbs light leaving most of the ocean dark. Smell and taste, or olfaction, are also limited, as molecules diffuse more slowly in water than in air, making olfactory senses less effective. Sound on the other hand can travel tens of miles, and some whale calls travel several hundreds of miles – across entire ocean basins.

High frequency sounds generally travel shorter distances and low frequency sounds can travel great distances. Cetaceans can intentionally vary their sounds by directionality, source level and frequency so as to communicate effectively. The SOFAR (sound fixing and ranging), or deep sound channel, results from the sound velocity profile of the deep ocean. A sound velocity minimum exists at the lower extreme, which is generally below the permanent thermocline (an area of sharp temperature drop). This 'minimum' creates a channel that acts as an axis or lens. When a portion of the source sound is trapped within this channel, it does not experience losses due to reflection or refraction from the ocean surface or bottom and can therefore travel very great distances with low transmission loss.

While a blue whale, which produces the lowest frequency sounds of any animal (below the hearing range of humans), in the North Atlantic can communicate with another blue whale in the Antarctic ocean, the function practicality of this communication is questionable. Humpback whales in the North Atlantic, on the other hand, that produce, in both song and other vocalizations, low- to mid-frequency sounds, may be able to utilize the characteristics of sound propagation to produce 'useful' communication with other humpbacks both near and far within a single ocean basin to serve a variety of purposes.

Marine scientists have discovered that the structure of humpback song has a predictable arrangement with a series of sounds repeated over time in patterns and each phrase repeated several times to comprise a 'theme'. A typical song is then made up of five to seven themes that are usually repeated in a sequential order. An individual song usually lasts eight to fifteen minutes, but may last thirty minutes, and then is repeated over and over in a song session that may last several hours.

A striking feature of humpback songs is that they gradually change or evolve over time. Each year, researchers like Jim Darling record different arrangements of sounds that form to create new phrases or themes. These changes are slowly incorporated into the song, while some older patterns are lost completely. According to researchers, the song changes as it is being sung and the changes in the song seem to happen in a collective way throughout the population. Jim has found that after a period of several years, the song is virtually unrecognizable from the original version although in some cases the song can evolve much faster, completely changing in just two years.

Marine scientists have also noted that despite the constantly changing nature of the song, all singers in a population sing essentially the same version at any one time. For example, all the singers in the North Pacific, including whales in Japan, Hawaii, Mexico and the Philippines, which are separated by thousands of miles, sing essentially the same version of a song at any one time. The songs of humpbacks may be similar across entire ocean basins like the North Pacific, yet different in separate oceans. Humpback whales in the South Pacific, for instance, have a different song from humpback whales in the North Pacific Ocean. The explanation for the collective change of the song, especially over such vast distances, is unknown.

Some years back a radical change took place in the song of humpback whales in the Pacific Ocean off Australia. Their song was replaced rapidly and completely by the song of the Australian west-coast population from the Indian Ocean, apparently as a result of the introduction of only a small number of 'foreign' singers. Such a revolutionary change suggests that novelty may stimulate alteration in humpback whale songs.

Photographs of the genital region and DNA sampling taken from skin samples can determine the gender of singers. According to some researchers, this evidence suggests that it is the male humpback whales that sing. Singers are usually, but not always, alone. There are several hypotheses made about the social function of humpback song. Some marine biologists believe the song is a sexual display to attract females and that the song changes as a result of female choice. Jim's research, however, suggests that the songs function to facilitate social interactions between adult males, with little or no evidence of female response. Like Jim, many marine biologists believe humpback song is some form of communication from male humpbacks during the breeding season. The song also likely broadcasts information about the individual singer, but what information is communicated and who the recipient is remain unknown.

While I was in Hawaii Jim invited me out on his research boat. He told me that a whale will sing until one of two things happen: they are joined by another lone adult male called a joiner, or the singer stops singing without any close approach by another male and then rushes to approach or join a passing group of whales, often with a potentially breeding female included within the group. When a humpback whale singer is approached and joined by another lone male, the interactions are usually relatively short, ranging from a single pass to rolling, lob-tailing or breaches by one or both animals. Often the pair splits again after a few minutes. One or the other may start singing again shortly after the interaction.

It was a calm morning when Jim and I left the Lahaina Harbor. Members of his research team were in other boats and they coordinated their movements to better understand the behaviour of the humpbacks. Soon after leaving the harbour Jim dropped the hydrophone overboard and we heard a whale singing louder than the background songs. Then a whale's back broke the surface of the water about 300 yards (275 m) away.

'There's the singer,' Jim told me with assurance. He started the engine and we drove in that direction before he turned the engine off. We heard the whale singing, only this time it was not with a hydrophone, but through the hull of the boat. I looked over the gunwale and in the clear waters I could see the whale below us. How Jim had managed to position us directly over that whale I don't know. Years of practice, I suppose.

'What we should see soon is another whale, a "joiner", joining this one,' Jim told me.

Sure enough, it wasn't long before another whale's back broke the surface of the water. It was a joiner. Both whales started moving off and within a mile other whales had joined in, or perhaps they had joined other whales. I have video footage of several whales moving fast alongside the boat. They were clearly on a mission.

'They've found a female, probably with an escort,' Jim told me as he searched ahead. The escort was a dominant male that had secured its position next to the female, effectively 'escorting' her while fending off challengers.

A few hundred yards later we witnessed these whales become a 'rowdy' group of several whales vying for the attention of the female with a young calf. This became a highly active surface group with head lunges, chin breaches, S-turns, sideswiping tail flukes and breaching. All Jim had to do was stick with the female because whatever action there was, it was all about her. After twenty minutes of high energy combat we were still with the female and her calf, and the dominant male was back at her side again.

The day's outing with Jim was a lesson on how much we can learn about the whales with a lot of determination and dedication.

Coming back to Bermuda after these trips to the Dominican Republic and Maui, my confidence levels were high. I was keen to get out on the water as soon and as much as possible and start to find out more about what the humpbacks were doing here in Bermuda.

Some calves are curious and approach without fear. Others seem nervous and hide behind their mothers while another might revel in showing off. Like humans, the whales have their own individual personalities.

CHAPTER FIVE

THE NORTHWARD MIGRATION

The first whales reported to leave the protective waters of the Caribbean for the feeding grounds are the females without calves. If they are pregnant, they need to start putting on the blubber as early as possible to sustain their calves next winter with their milk. The next wave to leave comprises the juvenile males, followed by the mature males and the last to leave are the females with calves. The young calves need as much time as possible in the protected warm waters of the Caribbean to put on weight from their mother's milk and develop before their long journey north. The timing of these waves of migrating whales does seem to match my own observations here, most noticeably the mothers and the young calves, which come by later in the season.

Many of the humpbacks begin their migration from the Caribbean up to their feeding grounds about the beginning to middle of March. This corresponds roughly with a ten-day to two-week travel time, depending on where they are going in the Gulf of Maine or Eastern Canada. If they are going to Labrador, Greenland, Iceland or Norway, the journey will be much, much longer. The first large wave of migrating whales arriving here in Bermuda consistently seems to be about the last week of March and continues until the last week of April.

The cards and posters I had distributed the previous year along South Shore asking the public to call with any whale sightings began to pay off. Phone-ins confirmed humpback whale sightings beginning at the end of December 2007 all the way through January and February to the main migration season in March and April. Some sightings were from fishermen out at Sally Tuckers and Challenger Bank, but most were from shore-based observations. Whenever possible, I searched the waters, whether I was sitting on the porch of our cottage or glancing out to the ocean while driving the car. I was surprised at how often I spotted the whales. I began to realize that there were far more whales in our waters during the winter than we thought.

At the end of February 2008 there was a reliable report in the media from a commercial fisherman who had seen an adult humpback, with green cargo netting wrapped around its head, to the northwest of the Bermuda platform near Chubb Heads. I had to wait four days for storms to pass by before getting a chance to go out on the ocean to look for the whale.

With friends and volunteers, I sailed out on the 44-foot (13.4 m) catamaran, *Windrush*, which was the perfect platform to search for the entangled whale. Despite the odds of finding the netting in the open ocean, shortly after passing by Chubb Heads on our way to Challenger Bank, while I was hoisted atop the mast, I spotted a bright green drift net that matched the fisherman's description. We hauled the heavy net on-board and, after spending the rest of the day looking for more nets or the entangled whale, we took the net to the Sea Shepherd Conservation Society's *Farley Mowat* that was then tied up in

Dockyard, waiting to begin its next campaign against the sealers in Eastern Canada. We placed the twisted mess on the *Farley Mowat*'s deck and as we unravelled the net it became obvious there was a large 'noose' where the whale must have been entangled. This large circle of net had rolled on to itself making a strong halter or bridle around the whale's head and body. Hanging from this loop of drift-netting was a mass of knots. Somehow the whale had managed to escape this hangman's noose. It was my first indication of how serious entrapments in ropes and nets are to whales and how they threaten the lives of the humpbacks and other whales.

Back in Maine, the research group Allied Whale at College of the Atlantic had been trying with some success to match our fluke IDs. Over the years the group had accumulated some long-term matches in their North Atlantic Humpback Whale Catalogue. In their data set, they found over 300 whales with a sighting history of 20-plus years, while there were only 19 with a history of 30-plus years. Two of those 30-plus whales had been photographed by me in Bermuda. The longest term match that Allied Whale has ever made is 36 years, HWC No. 0010, last sighted up in the Bay of Fundy in 2010.

But despite their best efforts, Allied Whale still hadn't identified Magical Whale, and eventually catalogued it as a new animal, No. 6090.

I realized by now how important our observations were to the process of understanding the humpbacks. I started making sure that my volunteers on the boat were always taking detailed notes of everything that happened during our encounters with the whales: the time, location, how often the whale surfaced to breathe, numbers and any other behaviours or activities; and environmental conditions such as tide, weather, waves and cloud cover. I asked them to write down anything that might be of some use later on for finding patterns in the humpbacks' behaviour. These sheets of information were duly typed up and formed an increasing databank of detailed observations. Besides filming the humpbacks, my primary objective, now that I knew how important the information could be, was to obtain the fluke ID of as many whales as possible.

Whenever I had a chance, I cycled along South Shore Road scouting for whales. If I found them I'd phone my volunteers to meet me at the ramp on Devonshire Bay. If Annabel was at home with Elsa I got them to climb on to the roof of the cottage to spot the whales and direct me to their position. If they were at home and it was a calm day, I'd quickly whisk them out on the boat to find the whales. Elsa did have a tendency to feel seasick and Annabel was apprehensive when we took the boat out of Devonshire Bay and through the reefs. For an adrenaline junkie who kitesurfs and only bothers windsurfing when it blows 25 knots or more, Annabel's anxiety going out in good weather reinforced my own misgivings at going out in much rougher conditions.

Finally, the hours I put in on the boat, jumping in and out of the water, paid off. Early one day we found whales near Chubb Heads. As two commercial whale-watching boats approached us we saw another whale breach about 2 miles (3 km) away in shallower water on the Bermuda platform. Rather than compete with the whale-watching boats, we headed off towards the whale that had breached. As had happened many times previously, we lost the whale in calm, 50-foot (15 m) deep water. Then we noticed the whale-watching boats heading towards us again, but this time the commercial boat captains weren't following us, they were shadowing four or five whales. On a hunch, I got into the water and immediately heard a whale singing. The singing was loud enough that I thought I must be almost on top of the source. As I swam around searching for the singer in the shallow water, the whale-watching boats continued to approach our position, still following the whales they were with. When the whale-watching boats accompanying the other whales reached us, the singing stopped, and six whales headed towards the point at Sally Tuckers where eleven fishing boats were moored or anchored. By now it was mid-afternoon and the commercial whale-watching boats headed back home, leaving us alone to observe the whales as they swam around the fishing boats where there were strong currents, with areas of calm bordering on bands of turbulent water from an up-welling. The fishermen and the whales were both feeding off the food web accumulating here.

(previos page) It is only by swimming with the whales that I can determine whether a whale is a yearling or not, and which whale is a calf's mother.

Attacks by orcas on the North Atlantic humpbacks have rarely been witnessed, yet a third of North Atlantic humpback whales have the characteristic rake-like scars from orca bites on their flukes or dorsal fins, and almost all of these have been inflicted when the whales were calves.

Then four of the whales departed, leaving a female and yearling. I was able to get some of my best video footage as the pair repeatedly approached me in 50 feet of water. Examining the footage, I could see that the fat 'calf' was clearly a yearling – about 26 feet (8 m) long, double the size it must have been a year earlier. The mother – and I can only assume it was the mother from their behaviour – was skinny and lethargic. The yearling was curious and approached me time and again, followed by its watchful mother. This yearling did headstands and cartwheels around me before we finally left two hours later to head back to Bermuda. By then the winds were fairly stiff with waves that made videography challenging.

While the yearling was very fat, the mother was emaciated, which might easily be explained if the mother were lactating and still feeding her offspring. If she had come up from the Caribbean, she probably hadn't eaten for some months. This is an indication that not all mothers abandon their calves during the first season of feeding up north, as I had read numerous times.

The mother of this calf had distinct orca bite marks, called rake marks, on her fluke. It must have been a frightening, horrific experience for her to have endured an attack by orcas, presumably when she was a calf. Good enough reason for a protective mother to remain with a yearling for the second migration northwards and then wean her offspring after reaching the feeding grounds the second summer up north.

Studies have shown about a third of North Atlantic humpback whales have the characteristic rake-like scars of orca bites on their flukes or dorsal fins and almost all of these are incurred when the whale was a calf. My footage shows this yearling acting like a playful puppy, almost goofy in its behaviour. It didn't seem to me that a young whale like this would have been capable of defending itself from an attack by a pod of agile orcas.

In the feeding grounds marine biologists have observed many mothers without their first-summer calves present, or calves and mothers that are much farther than a few miles apart. The mother could be feeding in the feeding grounds while her calf explores in the same approximate area but not in the immediate vicinity of its mother. As I had observed several times here in Bermuda, in whale terms, a calf being a few miles from its mother is no distance at all. Yet to the observer, the mother and calf might appear alone. This might explain the singing at the end of the season in the feeding grounds – a call to group mothers and calves and others together before heading south. But is it the males or the females doing the singing? While there is no evidence yet to suggest females do the singing, why would the males sing when they have no incentive? They aren't going to get 'lucky' up north, out of the breeding season. Marine scientists recorded some humpback whale songs off Eastern Canada and the Gulf of Maine throughout winter months in 2006 and 2007 so there is increasing evidence that some humpback whales are singing in northern locations in winter. Perhaps the young males are practising for when the time comes for them to head to the breeding grounds.

When I think back to that day, I wonder whether it was the fat, goofy yearling that was singing in shallow water, calling for its mother. Just like the other three- to five-month-old calves I'd seen in Bermuda, it might have been left alone in shallow water while the mother was presumably feeding on the edge of the Bermuda platform. Perhaps the yearling breached once to signal its position, as we had seen, and then sang until its mother, surrounded by other whales, came by. There is currently no scientific evidence that calves 'sing', although they certainly do make sounds. But I couldn't understand why an adult would have been in 50-foot (15 m) shallow water singing until a group of other adults joined up. Judging by the high volume, it seemed that the whale in shallow water very close to me was the one doing the singing, that the five other whales came to the singer, and there was definitely a yearling in amongst the group when they all joined up. Apart from the conventional wisdom that calves don't sing, the obvious choice for the singer was the yearling calf calling for its mother and her escorts. But as one whale acoustics expert remarked, that's a classic mistake to make and one to avoid, because high volume does not necessarily mean the closest whale. Song has a very intense source level and there is no way of knowing which whale is singing unless one has some sort of directional hydrophone. Perhaps it was one of the adults on the move doing the singing. After all, I wasn't in the water with the adults. Like many others, I still haven't figured this experience out.

I had seen similar situations where a calf, apparently alone in very shallow water on the Bermuda platform, began to lob-tail as we approached. Within five to ten minutes an adult whale, seemingly the mother, came to the calf. From these numerous observations it seems that the mothers might leave their calves in the safety of shallow waters of the Bermuda platform while they forage on the edges.

When I had seen yearlings in the Caribbean with their mothers some of them were curious and brave. Other yearlings were shy and reserved. Some mothers pushed their calves towards the swimmers and some mothers were protective. I had noticed that often the yearling calves that were shy had scarring on them, presumably from ropes or nets.

I used David's small boat out of Devonshire Bay less and less and started taking up offers of larger boats which would enable me to get out to Challenger Bank more often. I began to see that our encounters with humpbacks were more frequent on the outer edges, especially the south-west corner of the Bermuda platform near Sally Tuckers, 7 miles (11 km) offshore, and then all the way out on Challenger Bank. Bermuda isn't a mid-ocean seamount, it is an island, but we had immediate access to two authentic mid-ocean seamounts – Challenger and Argus banks, 15 and 25 miles (24 and 40 km) offshore respectively. This was a long haul on a small single-engine boat from Devonshire Bay, but on Bob Steinhoff's boat *Dom Perignon*, a twin-engine gas-fuelled Tiara 31 out of Somerset, it was a relatively short excursion.

Getting out on other people's boats, where I had no responsibility for provisioning or captaining, reduced my anxiety levels substantially. I could show up at the dock and a boat would be fuelled and ready to go with an experienced owner/captain. I just had to go for the ride and help find the whales, get in the water with them and get their fluke IDs.

The pectoral fins of a humpback whale are about a third of its body length -- roughly 15–16 feet. Humpbacks slap their pectoral fins on the water probably for one of the same reasons they breach -- to signal their position. Females lying on their backs and whacking the water with both fins often solicit aggressive behaviour from competing males.

While the yearling was very fat, its mother was emaciated, which might have been because she was lactating and still feeding her offspring.

At around this time another captain offered his boat. Getting out with Michael Smith on *Sea Slipper* increased my understanding of the whales' world for the simple reason that he had an easy-to-read GPS on the bridge where we could see our position in relation to the bathymetry – the depth of the water relative to sea level – of the ocean bottom. Now I was able to 'see' under the waves to the ocean bottom and look at what the humpbacks must also be able to identify. Mike's boat is a Monk 36 with a Perkins diesel engine compared to Bob Steinhoff's *Dom Perignon*, which is like a Maserati on steroids. Mike's boat is equivalent to an immaculate motor home, comfortable and slow, but with all the modern conveniences. From his dock it takes three hours to get to Challenger Bank, and therefore three hours to return, so our days are a minimum of six hours. Frequently we are on the water for twelve hours or more.

Less is more with Michael's *Sea Slipper*. By going slow we do not miss a lot as we progress south-west along the edge of the Bermuda platform, across the 4,000-foot (1,220) deep canyon between the Bermuda platform and Challenger Bank. Sitting comfortably on the top deck we divided the horizon into four quadrants and each crew member (usually there are four, sometimes five) searched his/her quadrant.

The marine chart on his easy-to-read GPS display informed us where we were relative to the ocean bottom with lines marking the 100-foot (30 m) contour lines. Coupled with the depth sounder, which functioned up to around 300 (91 m) feet, we were always aware of the features of the ocean bottom underneath us. We marked on the GPS every occasion we were within a hundred yards of a whale or whales. We didn't mark the GPS position if the whales were far away or if we thought it was the same whale some time later. The patterns of behaviour we thought we had observed seemed to be confirmed by these bathymetric reference points. If we saw whales showing their flukes repeatedly and diving on the edge of the seamount around 180-feet (55 m) deep, and coming up to the surface in more or less the same spot 8–12 minutes later, we surmised the whales were feeding on an up-welling over the seamount edge. As I learned to describe it from the scientists that I worked with at College of the Atlantic, this was 'behaviour commensurate with foraging'. More often than not, where we saw whales demonstrating behaviour commensurate with foraging, we also saw fishing boats demonstrating behaviour commensurate with fishing. Both whales and fishermen were on the up-welling on the edges of the Challenger Bank or Bermuda platform and they were all after the same thing: the abundance of marine life that thrives here, from plankton to small fish, tuna, wahoo, marlin and sharks. There appeared to be plentiful food to be had despite the scepticism of some marine scientists that there was enough food in Bermuda for whales to be feeding on.

I had begun to notice something else. Within the six- to eight-week period when we saw the most whales migrating by, we also saw the numbers of whales rise and fall in waves. There would be days within the peak migration when there were apparently scores of whales out there. Everywhere we looked we'd see whales spout, breach or lob-tail. A day later we might not see anything. It seemed that they had gone. If there was any pattern to this, the numbers increased leading up to the full moon and then decreased immediately after, before slowly building up again to the half-moon and then repeating the cycle. The problem was that we couldn't get out consistently on the days leading up to and on the day of the full moon or half-moon, but it was a tenuous pattern we noticed in 2007 that seemed to hold up in 2008.

I had purchased a good quality digital recorder and a decent hydrophone, but hadn't used the equipment since the beginning of 2007 because I didn't seem to get any sound. I realized after the first season that the hydrophone was defective and had it replaced. We were on Michael's boat the first time we used the new hydrophone. As soon as I turned the recorder on, we heard the whales singing. The effect was visceral. Listening to a whale sing in the ocean touches me to the core. It seems to have that effect on everyone. I've listened to the whales now hundreds of times and it still has a profound, primeval effect on me, almost as much as swimming with a whale. Now, with the hydrophone we could hear the whales, even if we couldn't see them. Knowing they were somewhere out there, we wouldn't call it a day until we located them.

Although we didn't have a directional finder on the hydrophone, we dropped the hydrophone periodically on the way out: on Sally Tuckers on the south-west corner of the Bermuda platform 7 miles (11 km) off Somerset, then halfway across the canyon and again on the north-west edge of Challenger Bank as we came up to the seamount. Consistently, the singing got louder as we approached Challenger.

By trial and error we observed that the singing was loudest in the middle of the Challenger platform. Challenger Bank is 10 miles by 15 miles (16 by 24 km) and is as flat as a billiard table except for the centre. Driving the boat around Challenger the depth sounder does not budge more than a foot or so off 170 feet (52 m) (depending on the tide) until we approach the middle of the bank, where it rises to 140 feet (43 m). It was here that we consistently heard the loudest singing. I guessed that the whale was singing on top of the cone in the middle of the seamount, but deep enough to be below the thermoclines (the areas where the temperature changes rapidly with depth).

Whenever I was in the water to find the singer, I could never locate it despite perfect visibility that might extend 100 feet (30 m) horizontally. Often the sound of the singing was so loud that my ribcage vibrated when I dived to any depth. Sound can travel far and can be focused or dispersed depending on pressure, temperature, salinity and bathymetry of the water and area. I am neither a sound engineer nor an acoustics expert, but it seems that singing below a thermocline and above a cone rising 30 feet (9 m) above a flat billiard-table plateau extending 5 to 7 miles (8 to 11 km) in any direction would be an effective way to propagate sound into the ocean. In fact, scientists know that once the sound is trapped between the top and bottom of the ocean it gradually begins to spread cylindrically, with sound radiating horizontally away from the source. Sound levels decrease less rapidly when spread from a cylinder, so this flat-topped seamount is an effective long-distance telephone booth. But we still didn't know who the singers were, who they were singing to, or why.

Looking for real-time patterns of coordinated whale behaviour related to whale song isn't easy – the whales are underwater and out of sight and may be spaced so far apart that it's impossible to observe these associations. But using the US Navy's antisubmarine listening system, the Sound Surveillance System (SOSUS), Dr Christopher Clark at Cornell University can track singing blue, fin, humpback and minke whales. Instead of tracking Soviet subs as they move through the Atlantic, the underwater microphones now provide a wealth of new data on whale song. Using SOSUS, Clark can move a cursor around a screen and listen in on different areas of the North Atlantic. If he hears a humpback whale singing, he can fix its location and position it in space and time using triangulation and observe cohorts of humpback singers moving coherently, despite being hundreds of miles apart. Clark reckons a whale off Newfoundland can hear a whale off Bermuda and has evidence that whales are communicating with each other over hundreds of miles of ocean. He believes that their singing is part of their social system and community. He can also listen to the collective migration of whales in large portions of the ocean basin.

When I met Chris Clarke at the biennial conference of the Society for Marine Mammology in October 2009, where I co-authored two poster-presentations with Dr Peter Stevick, I had a chance to quickly explain what I thought the singers in Bermuda were doing. He was entirely supportive: 'It's all happening in the middle of the ocean and marine biologists don't know it because it's easier to study whales closer to shore in their feeding and breeding grounds.'

Even if it wasn't whale song, that was music to my ears.

Finally, Mike and I found a two-day break in the weather that allowed us to spend the night out on *Sea Slipper* at Challenger Bank rather than commuting six hours there and back. When we reached the crown we dropped the hydrophone over and picked up the singing of humpbacks, loud and clear. We phoned an enthusiastic teacher at a school and patched in the singing humpbacks to his cell phone. He in turn connected the cell phone to speakers in his classroom and his students were able to listen to a humpback singing live 15 miles (24 km) offshore.

After looking for and recording whales singing throughout the day and evening all over the area to the southwest of Bermuda we headed off from the Bermuda Platform back to the crown of Challenger under a clear night sky. The further we boated into the darkness, away from the lights of Bermuda, the brighter the stars become. Phosphorescence streamed off the boat and as we crossed the canyon it became more pronounced until it looked as if we were in the midst of a snowstorm. It was eerily quiet so far from shore. At 2 a.m. we finally moored Sea Slipper on a fisherman's mooring on the crown at Challenger Bank some 20 miles (32 km) to the southwest of Bermuda in 160 feet (49 m) of water. We had inspected the line earlier in the day to make sure it was sturdy enough for us to tie on to. But sometime between then and tying up for the night, a fisherman had attached a rope to the line with baited shark hooks. As we pulled on the mooring line we hauled in a dead shark about seven feet long. It was a reminder to me that these waters are inhabited by large tiger sharks and hammerheads.

The night was absolutely still with an oily calm sea, perfect for staying overnight on the open ocean. We took recordings of the singing whales throughout the night until dawn when we slipped our moorings. In the morning we located a humpback under our boat and he remained there absolutely still, in the classic singer's position, pectoral fins outstretched and head slightly down. Was this the whale we had been recording?

One thing was becoming increasingly clear. The most activity happened around the singer. But these whales didn't seem to be the rowdy groups from the breeding grounds. Whenever we saw large groups of whales there was no overt high-energy jostling of males eager to dislodge a primary escort with a female. These whales seemed to be associated and moved in and out of each other's orbits as they careened around the Challenger Bank for hours, often breathing and diving simultaneously. Judging by their large and well-formed dorsal fins, with no typical male battle scars from competing to be the prime escort, it seemed there were multiple females with some males and many juveniles. I began wondering whether, rather than being simply a random aggregation of whales, this could be an association of known whales assembling into larger cohesive groups before heading north.

I was about to discard an unexceptional photograph of a humpback lob-tailing, when I noticed something extraordinary – the whale was defecating *(top)*.

Many male humpbacks have scarring on their chins, evidence of sparring with other males *(middle)*.

By consistently dropping the hydrophone overboard as we moved about, we were able to locate the singers *(lower)*.

Because we put a concerted effort into photographing their flukes, when we tallied all the individual humpback whale IDs for the 2008 season we counted sixty-eight individual fluke IDs. This was a lot more than the fifteen we had obtained the previous year. If we could keep this up each year, I hoped we would start finding some revealing patterns.

On occasions while crossing the canyon between Sally Tuckers and Challenger Bank we spotted other species including Cuvier's beaked whales. The Cuvier's were hard to discern, their backs only just breaking the surface and with no noticeable blow. We only saw them in the canyon on the calmest of days. They tended to breathe several times on the surface and then slipped underwater for about twenty minutes before surfacing. It was only by examining the photographs and sending them to experts that we deduced they were in fact Cuvier's beaked whales. I estimated the whales were about 15 to 20 feet (4.5 to 6 m) long. The Cuvier's beaked whale is difficult to distinguish at sea from the other beaked whales (such as the Mesoplodont whales, a genus of fourteen beaked species, also found in the area), some of which have been seen only a couple of times. If we did see them it was always a single whale. I guessed these occasional single sightings were males from their coloration pattern and numerous scars on their backs from intraspecies fighting with other males. Only once did we see several Cuvier's beaked whales together. When we put the hydrophone down to listen we could hear the clicking sound characteristic of beaked whales or dolphins.

One evening I was about to discard an unexceptional photograph of a humpback lob-tailing, when I happened to notice something extraordinary. When I examined the photograph more closely, I saw that I had captured the 'smoking gun' I'd been looking for – the whale was defecating. Given that it must have been four months or longer since the whale had left its feeding ground in the Caribbean, such defecation was one sure sign that whales do indeed feed on the mid-ocean seamounts. The photograph had provided another piece of the jigsaw puzzle, offering a further insight into the secret lives of the humpbacks.

(left page) While the Sea Shepherd's *Farley Mowat* was tied up in Dockyard, waiting to begin its next campaign against the sealers in Eastern Canada, Bermudian Laura Dakin gave me vegan cooking lessons once a week.

We began using a larger boat and going further out to sea in our search for the migrating humpbacks *(top row left)*.

This large circle of net had rolled onto itself making a strong halter or bridle around the whale's head and body. Somehow the whale had managed to escape this hangman's noose (top row right). *Farley Mowat* leaves for Canada, where the boat was confiscated by the Canadian authorities *(middle left)*.

If Elsa was at home and it was a calm day, I'd quickly whisk her out on the boat to find the whales *(bottom)*.

CHAPTER SIX

IN THE NORTHERN FEEDING GROUNDS

For the second consecutive summer, in 2008 Annabel, Elsa and I headed back to Nova Scotia. We rented a motorhome in Halifax and toured around the nearby province of Prince Edward Island (PEI). We didn't see any whales but I wanted to have a holiday that Annabel and Elsa could enjoy without having to devote the entire vacation to following my passion for the whales. Besides, Annabel was pregnant and expecting to deliver a month after we returned to Bermuda. PEI was ideal for taking it easy and the recreated village of Avalon where *Anne of Green Gables* is re-enacted was a hit with Elsa. We stayed there for two days. It seemed every other girl there, like Elsa and Anne of Green Gables, had red hair. The trade-off for sightseeing around idyllic but whale-less PEI was flying to the vast wildernesses of Newfoundland for the second half of our holidays.

Newfoundland has had a long history of whaling over many centuries and the remains of whales and whaling stations are to be found in 'outports' all along the coastline. Whalers in Newfoundland explored uncharted territory in their insatiable search for whales. Like many of the other New World frontiers, it was the whalers who founded these outer edges of civilization. We began our own Newfoundland research by travelling immediately from St John's international airport to the nearby Witless Bay Ecological Reserve. Within minutes of arriving, in the evening's slanting sunlight, we could pick out dozens of whale spouts. The next morning we were out on the water in a small Zodiac – a type of inflatable boat with outboard motor – surrounded by humpback, fin and minke whales. It was a whale enthusiast's dream come true – lots of whales, unperturbed by our presence, feeding on the spawning capelin, a type of small forage fish.

There were also plenty of feeding puffins and other exotic seabirds as well. The puffins were so fat they could barely take off from the surface of the water. Whales and birds were everywhere in the feeding frenzy. It amazed me that the humpbacks, the giant fins (second only in size to the blue whale) and the small but fast minke whales were in such close proximity. It almost seemed as if they were cooperatively corralling their prey.

On the second day we noticed a mother humpback and calf. The calf listed to one side every time it surfaced. When the calf spouted right beside the Zodiac we noticed an open wound just behind the base of the pectoral fin. From the position of the wound it didn't seem likely to be a result of a propeller cut or an entanglement. That summer there were anecdotal reports of sightings of orcas attacking a humpback off the Newfoundland coast. Documented orca attacks on humpback calves are rare in the North Atlantic, but witnessing an injured animal such as this was an indication that orcas do attack young humpbacks in these waters. The wound, although fresh, did not seem to be severe, and I am sure the humpback calf survived the injury. This wounded calf was a first-hand indication why the humpbacks would be motivated to group together into larger social units for protection against the orcas as they approach their northern feeding grounds.

We walked along Newfoundland's East Coast Trail where we lay down in an open meadow on the headland and counted several groups of humpback, fin and minke whales.

How would travelling together protect the humpback calves against orca attacks? Baleen whales' strategies for defence fall into two categories. The 'flight' species, such as the blue, fin and minke whales, use their high-speed capabilities to minimize the risk of predation by avoiding encounters with predators and by quickly escaping after an encounter. The 'fight' species such as humpbacks and grey whales are not fast enough to flee so they retaliate and fight back. Fight species generally use calving areas where they are in numbers to provide favourable conditions for protecting calves from orcas. But there are no safe corridors in the middle of the Atlantic, unless the humpbacks travel together. Attacks by orcas on humpbacks are rarely witnessed and some scientists do not consider humpbacks to be a main prey choice for orcas. But, with a third of the North Atlantic humpbacks bearing the characteristic orca rake marks, it is clear that a substantial proportion of humpbacks are being targeted by orcas. And the calves bearing the scars are the ones that survived.

Long-term photo-identification studies of humpback whales in the Gulf of Maine reveal that virtually all scars consistent with orca teeth are already present on individuals when they are first identified as juveniles. I've never heard of orca sightings in Bermuda and I have yet to see a new humpback calf with rake marks, an indication that they are not being targeted south of here. Because orca sightings are rare on both the calving and feeding grounds of the North Atlantic population, it is thought probable that the attacks occur away from the feeding grounds, somewhere along the migratory corridor during the calves' first migration.

The apparent high incidence of orca scarring may not be so much a reflection of high rates of deaths caused by orcas but more an indication of the regularity of probing confrontations where the orcas seek a weak potential target. Even a failed attack by orcas must be a horrific experience for mother and calf. From witnessed accounts of lethal orca attacks on grey whale calves, the process is a long one before the victim finally dies, usually from drowning or excessive blood loss. For the specific humpback calf embroiled in an orca attack resulting in its death, it is a tragedy. But, looking at this from a Darwinian perspective of survival of the fittest, in theory the orcas' relentless probing and culling of the weaker individuals will result in a stronger, more robust humpback species as a whole.

Having seen humpback calves' behaviour underwater, I would conclude they are no match against orcas. The calves' survival against orcas, and this includes the yearlings in my opinion, depends on effective protective actions taken by their mothers and possibly other individuals connected to the mother/calf. Even if the proportion of lethal attacks on humpback calves is low, the frequency of unsuccessful attacks provides plenty of reason for humpbacks to travel in protective convoys. These humpbacks associated within a protective group may not be obvious to the casual observer, they could be miles apart. But, as I have witnessed around Bermuda, a lob-tailing whale will often soon have another whale or two by its side where there had been no previous evidence of other whales being in the vicinity.

Even if present rates of lethal predation are low, they may not reflect higher predation levels in the evolutionary past, when successful anti-predator strategies evolved. Despite the low-risk of mortality, just the presence of predators would have far-reaching effects on a prey species' behaviour. Migrating north into orca territory within a defensive group of humpbacks prepared to protect their young would reinforce these evolutionary anti-predator traits as the survivors using those strategies are the ones that go on to reproduce.

The humpbacks' ability to fight and defend themselves is enhanced by the high densities of embedded acorn barnacles found on their fins and flukes. The humpback whales derive protective advantages from their barnacles that outweigh their cost in reduced speed resulting from the increased drag. Humpbacks' skin may even be specialized to tolerate or encourage barnacle attachment. On the other hand, whales such as blues, fins and minkes may have evolved anti-fouling mechanisms to prevent such infestation. Scientists have provided evidence that larval settlement of the barnacle results from a chemical cue from the humpback whales' tissue, favouring the evolution of a mutually beneficial relationship between barnacles and humpback whales due to the barnacles' usefulness as defensive weapons in interactions with killer whales. The razor-sharp barnacles on the leading edge of a humpback's 16-foot (5 m) long pectoral fins are akin to brandishing a sword in combat. Likewise, the barnacles accumulated on the swollen tips of their heavy flukes act as the sharp spikes on a club-like mace.

Because of the apparent lack of sightings of orca attacks on humpbacks in the North Atlantic, at least around the Gulf of Maine and Eastern Canada, it might be assumed that once the whales are within striking distance of their feeding grounds they are relatively safe. Humpbacks are probably as likely to encounter orcas at the feeding grounds, but once there the humpbacks are relatively safe, being in numbers. Although they disperse to some degree once there, they would still be close enough to defend each other from orca attacks. There seems to be increasing evidence that singing in the feeding grounds, especially further north, occurs in the autumn. This could be prior to the whales leaving their safer feeding grounds to cross areas further offshore where they might encounter orcas on their southward migration.

Between outings on the water in the Zodiac in Witless Bay, we walked along the East Coast Trail. The first day we were shadowed for some hours just a stone's throw out to sea by a mother humpback and her calf. When we reached another bay we lay down in an open meadow on the headland and counted several groups of humpback, fin and minke whales.

From Witless Bay we proceeded north to Trinity on the Bonavista Peninsula overlooking Trinity Bay, where we stayed with Dr Peter Beamish for four days. We went out twice a day on Zodiacs where we not only observed humpback, fin and minke whales at close quarters and in abundance, but also sperm whales feeding further out over a deep-water canyon. Elsa was awestruck at the small icebergs we encountered drifting along the rocky shore.

What was particularly impressive to me was not just the number and variety of whales, but the dense population of capelin. The waters were dark with their biomass. On several occasions we found them spawning on the pebbled beaches,

The waters were dark with the biomass of dense populations of capelin. They spawned on the pebbled beaches and returned to the ocean to die or to be consumed by the foraging whales or the many bald eagles and other oceanic birds.

(next page) The best part of the Newfoundland trip for Elsa was seeing and touching icebergs -- something we never see in Bermuda!

returning to the ocean to die or to be consumed by the foraging whales or the many bald eagles and other oceanic birds. On one beach we saw a couple of dozen bald eagles feeding off the stranded capelin. For the past few years, scientists have observed that the capelin population has been changing distribution, likely moving further offshore, followed by more humpbacks to these offshore regions.

A generation ago, the waters here were dark with the biomass of cod. Children in Newfoundland could catch cod simply by dipping a basket into the ocean. Now Canadian research vessels scour the seas looking in vain for a single school of cod in what was once the world's richest fishery. We found the occasional rack of half-a-dozen cod set out to dry but the rich cod grounds have been so depleted that local fishermen nowadays are only permitted to fish for a couple of cod per day during regulated days in the summer and fall, called the Recreational Cod Fishery. The destruction of the Grand Banks cod stands out as one of the biggest fisheries disasters of all time. Opinions on who is to blame and who contributed varies depending on who you ask. The fishermen blame the scientists. According to the scientists, they warned fishermen and politicians of the cod collapse for years before it actually happened, and say their scientific advice went unheeded. Beyond the Canadian 200-mile (320 km) limit (the Exclusive Economic Zone or EEZ), any country can come in and fish cod. It wasn't the local dory fishermen who caused the cod fishery to collapse: it was the Canadian and international 'fish-factory' ships that plundered the cod population. It's still a very controversial topic, but the fishermen too should be blamed for misreporting catches, as well as the politicians who ignored the scientists' advice. Whoever's viewpoint you side with, they all played a key role in the collapse. Too late, the Canadian government banned fishing on the Grand Banks in 1992, when their scientists discovered there were nearly no adult cod left. Canada blamed Spanish fishermen, the seals, the weather and even the whales. But the real damage was done by years of 'safe' catches that scientists belatedly realized were just the opposite.

Some Japanese scientists continue to insist that whales reduce fish stocks, leaving less for humans. Japan has even suggested that whales consume six times

the world's commercial fish catch. But the seas were teeming with both fish and whales for millennia – until humans came along. The key change was the arrival of steam power, which allowed trawlers to plunder the open oceans.

We did see racks of capelin set out to dry. Decades ago instead of racks lined with the tiny capelin, they would have been laden with the much larger cod. Now the capelins' primary predator, the cod, have gone and capelin numbers have increased dramatically, which must be good news for the whales. The humpbacks move along the Newfoundland coast in waves, following the spawning capelin. The only time we observed the humpbacks breaching was when they were off the headlands of the bays, an effective location if signalling their position to other whales. If many whales breach off a headland it might be an obvious signal to other distant whales that there were many whales in the vicinity and that food was available.

I heard from many Newfoundlanders that humpback whales come in very close to the rocky shoreline to feed on the spawning capelin. Often the humpbacks seem to be almost on the pebbled beaches and this probably accounts for the white scars along the lower right jaws of the humpbacks. The scars look like severe abrasions resulting from the scraping of the jaw on a rough shore. Like humans, who are left or right-handed, the majority of humpbacks seem to favour one side: usually the scarring is on the right side of their jaws – in fact, all of the scarring I have witnessed underwater has been on the right. There isn't any reason I can determine why humpbacks would be 'right-jawed' any more than scientists have determined why humans are predominantly right-handed. Unlike other mammals, which do not demonstrate a consistent 'sidedness', human and humpback brains seem to be hard-wired similarly.

Having observed whales in New Zealand, Vancouver Island in British Columbia, off the East Coast of the United States, Massachusetts, Maine, New Brunswick, Nova Scotia, the Caribbean, Hawaii, Antarctica, Spitsbergen and Norway, I can vouch for the fact that whale watching in Newfoundland is as good, if not better, than anywhere else in the world. There are not many places where you can easily see humpbacks, fins, minkes, sperms and dolphins in a morning outing. And there can't be many places where you can be so close to a fin whale that you can observe, from land, the asymmetrical white patch on the lower right jaw.

A week after returning to Bermuda, Annabel and I were free-diving a few miles off Bermuda in the shallow reefs ringing the perimeter of the Bermuda platform. Annabel climbed into the boat and said she thought her waters had broken. It had been difficult to tell while swimming in the warm ocean but it soon became apparent that she was right. We told Elsa that her baby sister was about to arrive and we motored back to shore. Four hours later, Somers was born, a month earlier than expected. She had announced her arrival in a very appropriate way – while we were in the ocean we love so much.

Witless Bay Ecological Reserve is a whale enthusiast's dream come true -- lots of whales, puffins and other exotic seabirds unperturbed by our presence, feeding on the spawning capelin.

CHAPTER SEVEN

SLEEPERS AND MATING

For the second time, I flew down to the Dominican Republic to spend a week on a live-aboard boat. It wasn't the same as being on a boat with my own crew, but it did give me the opportunity to film very young calves with their mothers.

The trip started off badly with winds of 35–45 knots and seas as high as 16 feet (5 m). We were scheduled to depart from Puerto Plata on Saturday evening, arriving at Silver Bank early on Sunday morning. The waves at the entrance to the marina were too high and, unable to depart, we sat in the harbour until Tuesday. The crossing, when we finally embarked, was done at 2 knots, taking us 30 hours to reach our anchorage. It took another 20 hours to get to our moorings off the reef line.

The upside to the stormy weather was the enormous number of whales that seemed to have congregated closer to the protective ring of reefs where we were moored. It seemed with the winds coming predominantly from the south-east, the whales, especially those with young calves, were assembled close to the breakers for protection from the waves. They were also active at the surface during the numerous squalls that came through, which provided plenty of opportunity to photograph breaches and lob-tailing. Humpback whales usually communicate through vocal signals. When the wind and waves are high, the humpbacks seem to breach, lob-tail or slap their pectoral fins more often. Surface-generated sounds such as breaching are like loud explosions and have their energy distributed over a greater frequency range, making them less likely to become lost in wind-generated noise. The loud splashes of bodies, flukes and pectoral fins hitting the surface could be a more effective way to communicate in such noisy background conditions.

Because we only had a couple of days on Silver Bank, and because the water was so stirred up, I obtained little usable underwater footage. When we returned to Puerto Plata, there was a last-minute cancellation on the following trip, and I opted to take the spot. That second week on Silver Bank I obtained plenty of underwater footage of young, curious calves that danced around me before retreating to the protective custody of their mothers. On one occasion a calf below me opened its mouth and engulfed a mouthful of water before rising to the surface beside me. Its mother followed and as I hung motionless in the water her fluke passed within inches of my head. She had deliberately kept her lethal, barnacle-encrusted tail immobile as she swam by. A little twitch of that mace-like fluke would have seriously injured me. So much larger than us, the humpbacks are acutely aware of their extremities and seem to go out of their way to avoid physical contact with humans, as I had observed over and over with Magical Whale.

On another occasion a young female surfaced with her male escort. She reached out with her pectoral fin as if to make sure I kept my distance, but then deftly ducked the fin under my body. She deliberately and carefully avoided

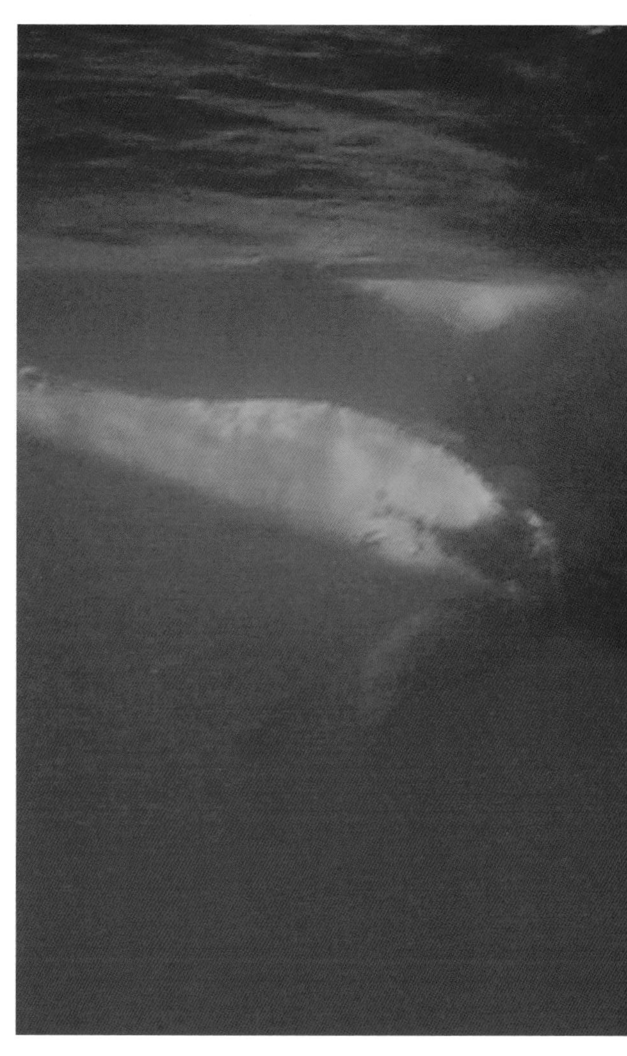

Mothers often swam directly at me, as if teaching their calves about the world around them. Other mothers were more protective, placing themselves between me and their calves.

touching me, which would have been a significant blow from my point of view. This youngish-looking female had a deep gouge at the base of her pectoral fin, a scar that matched the same kind of wound we had noticed on the calf in Newfoundland. I wondered if this was the result of an orca deliberately trying to impair the swimming ability of a calf with a bite to the base of the pectoral fin. It was the second time I'd seen a scar like this underwater.

My most interesting insights into the whales' lives were based on these underwater observations. Trying to study whales from the surface is akin to studying elephants by examining the tips of their trunks underwater when they come to a waterhole to drink. Being in the water with the whales provides a window into their lives that is just not possible from surface observations.

Occasionally we found 'sleepers' underwater, whales that seemed to be sleeping below the surface, often head-to-head or close beside each other. They almost always included one smaller male with lots of scars, and a larger almost unmarked female. Whales generally 'sleep' at the surface while 'logging' – when they drift or swim very slowly, resembling a log, hence the term. Because whales are voluntary breathers, unlike humans who breathe automatically, they sleep with half their brain while the other half is alert. I've never seen humpbacks mate, but to me these sleepers looked like whales that have either just mated or, more likely, are about to mate. During our first week we were lucky enough to have a couple of pairs of sleepers within a hundred yards of our boat and we were able to dive down to them over a relatively long period of time. Once again I could hear one or both of the sleepers 'chirrup' at a very low volume that was impossible to hear at the surface but was just audible when I dove down to them. It is possible that they were also making vocalizations below the audible limit of humans.

Dan Henriksen, another diver on the trip, took a photo of an aroused male – irrefutable evidence that the male of a pair of sleepers is obviously keen to mate. One can get an idea of sizes and proportions by keeping in mind that the flukes are 12 feet (3.7 m) across, and the pectoral fin is about 15–16 feet (4.6–4.9 m) long. The penis is not dwarfed by either fluke or pectoral fin and appears to be at least 6 feet (1.8 m) long if not more.

Leaving Silver Bank on Friday morning for the crossing back to the marina at Puerto Plata, I watched the whales until I could no longer see them. This uninhabited corner of the world, awash with humpbacks, fills a special place in my heart.

(left) When the wind and waves are high, the humpbacks seem to breach, lob-tail or slap their pectoral fins more often.

(this page) She reached out with her sixteen-foot pectoral fin as if to make sure I kept my distance, but then deftly ducked the fin under my body. The razor-sharp barnacles on the leading edge of a humpback's fin make it as sharp as a well-honed sword.

An aroused male keen to mate *(lower left)*.

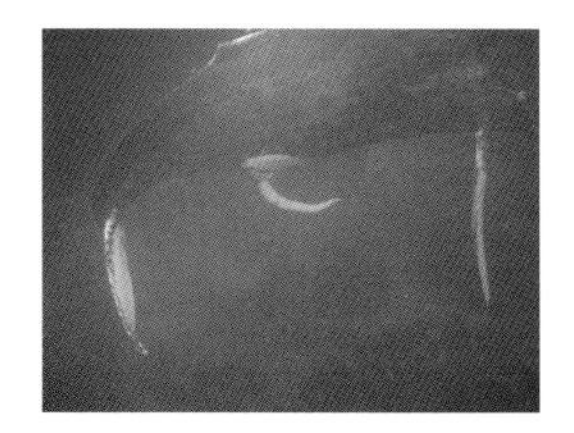

A calf plays, dancing around in the elemental joy of being alive (top two rows).

The calf opened its mouth wide to engulf a mouthful of water before rising to the surface beside me *(lower left)*.

I hung motionless in the water as her fluke passed within inches of my head. She had deliberately kept her lethal, barnacle-encrusted tail immobile as she swam by. A twitch of that mace-like fluke could have decapitated me.

The aggressive behaviour of the males as they batter each other in attempts to win the coveted position of escort next to a female often leads to bloody tubercles, backs and dorsal fins. This male has launched itself on top of a competing male.

The power of these whales is evident when they launch their bodies out of the water to sideswipe another whale.

(right) Lifting its tail high out of the water to plant it firmly on top of a pursuing whale is one of many strategies the males use to rid themselves of competitors.

(left) Another strategy to lose a competitor is for a whale to stop in its tracks so the pursuing whale ends up in front.

Several whales battle it out as they use their heads and tails to dissuade other suitors from approaching the female.

This male prepares its primary weapon, its twelve-foot fluke, ready to lash out in any direction.

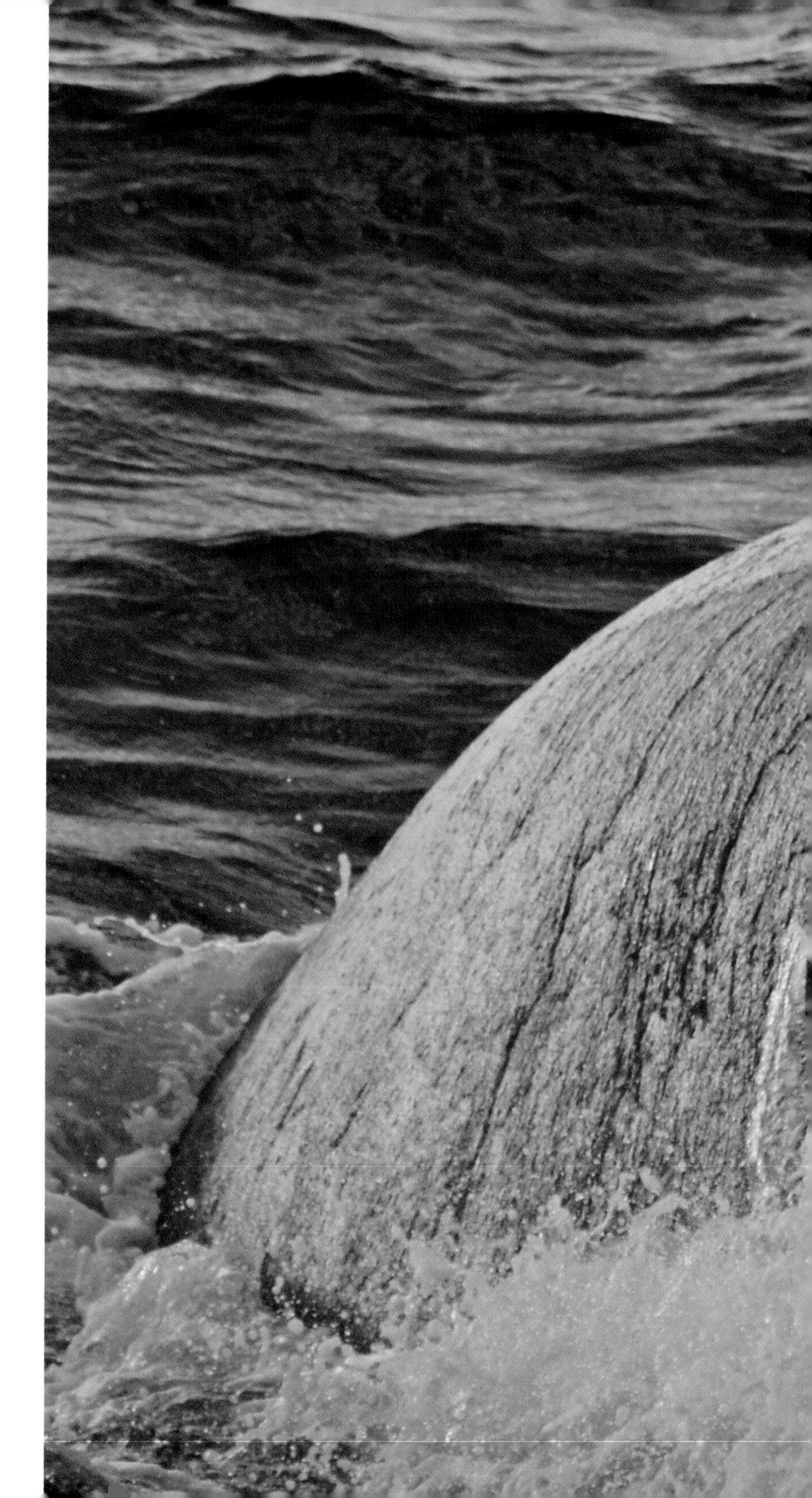

Males will partially fill their throats with water to lend them more weight and force when they batter an opponent.

CHAPTER EIGHT

THE 'GRANDMOTHER HYPOTHESIS'

On 15 January 2009, while looking casually out from our cottage, I spotted two humpback whales spouting about 300 yards (275 m) off the breakers of Grape Bay Beach while moving west. They spouted once again and then disappeared. I picked Elsa up from school and drove to a high point on the South Shore and after fifteen minutes, two whales surfaced about 50 yards (45 m) from the breakers, almost on the shoreline in 70 feet (21 m) of water. One whale did a small lob-tail and then they vanished. With the extensive views afforded from that vantage point and given the calm conditions and slanting sunlight, I knew I would see them when they resurfaced and blew. It was almost half-an-hour later when the two whales spouted again, only 200 yards (182 m) from where I had observed them before and still in shallow water. They dove again and then disappeared. I waited twenty minutes, but didn't see them resurface. I left as it became dark.

I didn't know what these whales were doing for half-an-hour in almost the same spot in this relatively shallow water. They could have been singing, resting or feeding. My best guess, based on the smaller spout, is that this was a mother with a calf, perhaps a yearling. They were either resting or perhaps the calf was feeding from its mother in the protective, shallow water.

One of the most extensive studies of humpbacks took place in the North Atlantic Ocean. This three-year study revealed that while the male to female ratio of humpbacks is more or less one-to-one in their northern feeding grounds, it is four-to-one in the southern breeding grounds. So where are the missing females? I think that they must be on these productive mid-ocean seamounts between the breeding and feeding grounds. If a female humpback whale is too young or too old to breed, there is no reason to go all the way down to the Caribbean where there is little access to food and where they are harassed by overzealous, testosterone-driven males. Even the young male humpbacks, too small to compete with larger males for a female's attention, seem to join the party down south hoping that they might perchance get lucky. It seems very likely to me that the younger and older females perhaps avoid being harassed in this 'bar scene' down south and instead hang around these mid-ocean seamounts where they can opportunistically feed off the krill, zooplankton and small fish that rise towards the surface in the up-welling currents.

Going through Bermuda's archives I came across several references to humpbacks being in Bermuda waters during the winter before whaling really took hold, although the greatest numbers occurred in the March to June period. Perhaps the humpbacks are re-colonizing breeding grounds they used centuries ago, before they were almost exterminated.

Dr Hal Whitehead from Dalhousie University in Halifax, Nova Scotia, had sailed from Halifax to the Caribbean the previous February to find sperm whales, with several postgraduate students on his sailboat. They towed a hydrophone behind the sailboat the entire way, but listened and recorded

Often we saw over a dozen whales surfacing in a co-ordinated way every few minutes, blowing several times before fluking together. They careen around the crown of Challenger Bank, sometimes for days at a time.

observations every thirty minutes. As soon as they reached the shelf break south of Nova Scotia, they started hearing a lot of humpbacks and sperm whales. On 45 per cent of those occasions, they heard humpbacks singing. If the humpbacks were singing, this means there were males around, unless the females sing, but, as mentioned, whale experts do not seem to think that is the case. Perhaps there are enough young and old females overwintering on the mid-ocean seamounts that some of the males remain there, too. The singers could be young males unable to compete in the breeding grounds so they stay further north and 'practise' their songs. Or maybe this is indeed evidence of a population range shift, the humpback whale numbers increasing to the extent that southern breeding grounds are too small to accommodate all of the whales, forcing some to stay in more northern areas.

You would think that when the humpbacks are singing near the mid-ocean seamounts, which seems to be the case, that the orcas would hear them and come to attack, but they do not if there are no young calves around, or if the humpbacks are loosely grouped together in larger social units for protection.

The 'grandmother hypothesis' explains why menopause, rare in mammal species, arose in human evolution and how a long post-fertile period, up to one-third of a female's life-span, can be an evolutionary advantage. The 'grandmother hypothesis' states that menopause, which stops a female's fertility well before the end of her lifespan, evolved to benefit a social group, because grandmothers play an important role in helping to care for their daughter's offspring. Cetaceans are the only other mammals where females have a post-reproductive lifespan comparable to humans. I don't think there is hard scientific evidence to this effect, but perhaps older female humpbacks also reach menopause and no longer breed, and become available to play the role of 'grandmother', looking after the young in the middle of the ocean. Although there is also no hard scientific evidence of females singing yet, no one has identified the gender of the mid-ocean singers in the North Atlantic. Perhaps we should not rule out the possibility that 'grandmother' humpbacks are singing on the crown of Challenger Bank, gathering family members to travel to their feeding grounds together.

The three-year study of the North Atlantic humpback whales referred to above also revealed that there was a general mêlée of whales in the breeding grounds down in the Caribbean. In other words there was no rhyme or reason for a whale to be in a certain location at any point in time. It's like a big party scene with males chasing females and trying to mate. Despite mooring in the same location year after year, the swim-with-the-whale tour operators on the Silver Bank don't see the same whales over and over again. However, as any whale-watch tour operator up north knows, a lot of humpbacks do maintain site fidelity to their feeding grounds. The question is, do the same whales meet up in their habitual feeding grounds or do the humpback whales assemble into loose social units on the mid-ocean seamounts before travelling together up north?

With 30 per cent of the North Atlantic humpback whales having bite marks on their flukes, pectoral or dorsal fins inflicted when they were calves, there is a very real incentive for the whales to gather into large protective social units, mid-ocean, before continuing their migratory route north of Bermuda, and into orca territory. If this is the case, then the humpbacks' social ties and responsibilities to each other are far stronger than has previously been supposed, and are more intricate than just being a matter of feeding and breeding. The singers, males or females, could have social duties to their kin or species and are gathering families into groups to continue migrating together.

Not only does it seem that Bermuda is an important stop en route to the northern feeding grounds, but it also seems that the whales may be staying here throughout the winter months. This is not to say there is a resident population, but the whales are often found around these mid-ocean seamounts, where food is much more abundant than we previously thought.

Fishermen told me they had never before seen as many bait fish, whales or large predator fish (tuna, sharks and wahoo) as in the winter of 2008–9. One fisherman told me he had seen whales almost every day on Challenger or Sally Tuckers for thirty days, from half-way through December to mid-January.

There was obviously a correlation between the large numbers of humpbacks seen this winter and the considerable amount of finger-sized shrimp or krill. Then, the first week in January a fisherman saw what he described as a 12- to 15-foot (3.7 to 4.6 m) humpback whale with a larger whale. At 12 to 15 feet this could only have been have been a baby humpback whale a couple of weeks old. The identification was reliable, made by a trusted source, based on a calf only 50 yards (46 m) from the boat. This is when calves are usually born in the Caribbean, so this calf could not have migrated up to Bermuda so early in the season and was likely to have been born in an area much closer, perhaps on one of the mid-ocean seamounts. It could have been a premature birth, its mother not making it all the way down to the Caribbean in time. Or, a newborn calf in Bermuda waters could be an indication of a resurgent humpback whale population re-colonizing old breeding grounds. Or it might be an indication of overcrowding in the Caribbean caused by an expanding population.

This same fisherman that described the humpback calf also collected contents from the stomach of a tuna caught on Challenger Bank around the same time and gave it to me to take to the Bermuda Aquarium for analysis. The 5 pounds (2.3 kg) of frozen stomach content were defrosted and examined by Dr Wolfgang Sterrer and Chris Flook. The bulk of the content consisted of fish, most of which was digested beyond species identification. Among those that could be identified were a number of anchovies and pilchards, which are commonly used as chum bait. Possibly also originating from bait was one specimen each of halfbeak and ocean robin, the latter often being used as live bait. Other fish species included: one small Sargassum triggerfish, which is usually associated with floating Sargassum seaweed, one lantern fish and five bristle mouth. The last two, which are deep-water species, could have been caught by the tuna diving, or at night when these fish come to the surface. There was also one small teuthoid squid and at least thirty red decapod shrimps, apparently all belonging to the same species.

This analysis was further indication of the different levels of food available in up-wellings on the seamounts.

One Saturday morning early in January 2009, I walked onto our porch and, looking out to sea, spotted the arching back of a humpback not 200 yards (183 m) from the breakers. It was a pair of whales and it seemed that one had a smaller blow than the other. Over the next weeks I had multiple sightings of whales phoned in from along South Shore in waters no deeper than 100 feet (30 m). Often these anecdotal reports indicated a pair of whales with the blow of one being smaller than the other. There were also groups of whales sighted and these tended to be surface active with lob-tailing and breaching.

Another pattern that we observed as the spring season progressed and we spent more time out on the water was that we were consistently hearing the loudest whale song on the crown of Challenger Bank.

One day in March we were right on the crown of Challenger and, with the hydrophone overboard, we could hear singing that was very loud and obviously very close. A whale surfaced at intervals with a curious splashing that I had never seen before. While the whale surfaced, the singing continued uninterrupted. We followed this whale with the peculiar slapping of its pectoral fins against its body and surmised that it was injured. It seemed to be entangled in a net or line with its pectoral fins pinned to its sides. Every time it surfaced to blow it whacked its pectoral fins audibly against its sides. What we presumed to be the singer joined and escorted the crippled whale. When I examined the photographs at home that evening they showed the pectoral fins high up on the whale's back in an unnatural position and yet there was no obvious indication of an entanglement with ropes over the back of the whale. I guessed that this whale was not entangled at the time we saw it.

Having now seen other entanglements pinning the pectoral fins to the sides of a whale, I believe this one must have been entangled for a lengthy period of time. I think that when it had freed itself of the tangle of ropes, it had been left crippled, its muscles and ligaments at the base of its pectoral fins no longer able to extend its pectoral fins outwards. As a humpback arches its back to dive its natural tendency is to extend the pectoral fins outwards and angle them to help

We found a calf with an entanglement around its head in shallow 30-foot water. The mother was emaciated. She had had a large, visible chunk, still hanging by a strip of skin, taken out of the trailing edge of her right fluke.

it dive. All this handicapped whale could do was to smack its fins audibly against its sides with a peculiar splash and slapping sound. Perhaps the singer was escorting this crippled whale. Whether this was a short-term injury, it being recently freed and not yet regaining its strength, or a more permanent disability I do not know. Unfortunately, we did not obtain a fluke ID of this whale, which would have helped us to compare its behaviour in later years.

A few days later I was in the water with what seemed like four immature whales, judging from their smaller size, longer than 26 feet (8 m) but less than 45 feet (14 m). While it would be almost impossible to tell these were smaller whales based on observing their backs from the surface, it becomes more apparent underwater. They moved in tight formation, never more than a pectoral fin away from each other. One of them was a flirt and exposed her underside to me, revealing she was a female. Curious about me, they swam very close multiple times, always tightly packed together. One seemed to be a young male judging by the long white scars on his dorsal fin and along his back and by his possessive behaviour.

It ended up being a very long day in and out of the water from shortly after midday to six in the evening. Most of the whales were females or of unknown gender. These whales were well fed. They did not show any tell-tale signs, the bony vertebrae along the top of their backs for example, of having starved for the past few months, as do the whales that migrate up here later in the season from the Caribbean. They seemed to be healthy, mostly young females perhaps with just one young male, all overwintering in the middle of the ocean.

Some days later I received an email from Dr Peter Stevick with a match to one of my underwater images of a female's flukes as she passed by me:

That beautiful underwater photograph that you recently sent of your whale #0146 has some very distinctive markings. It jumped out at me yesterday as I was going through the collection for other reasons. It is NAHWC#8732 – Pogo. We have Gulf of Maine records of that whale from 2002, 2006, 2007, 2008, and now Bermuda 2009.

That would make our young female seven years old, at least. She was the bravest of the four whales and it was she who consistently came close to me, showing her belly (and her genitals) and the ventral side of her fluke that we used to ID her. If she wasn't seen before 2002, and if she keeps showing up consistently in Maine, I'd take a guess that she is a young female ready to go south next year.

One of the patterns revealed in the matching of 'our' Bermuda fluke IDs with Allied Whales's North Atlantic Humpback Whale Catalogue was the high number of matches to Gulf of Maine whales. Looking at a map, the direct route from the Silver Bank to Maine isn't through Bermuda. Bermuda is in a direct line to the easternmost edge of Newfoundland, and in line with Labrador, Greenland and Iceland. It puzzled me why the Gulf of Maine whales didn't travel closer to the continental shelf where they could catch a ride with the Gulf Stream. If the seamounts were being used as gathering points because of the good acoustics, I wondered if the higher proportion of Maine whales coming further out to sea to Bermuda to assemble was because of the noise pollution along the eastern seaboard.

But these observations still don't explain why we hear humpbacks singing 24–7 here in Bermuda as late as May, long after the breeding season is over. If it is the males singing, the singing in Bermuda has nothing to do with breeding, unless they are re-colonizing old breeding grounds, which would explain why they are singing. With an eleven-month gestation period no female is going to want to breed in April/May, so late in the season, unless there's a shift in the breeding time as well as the location. Surely a male humpback is too smart to waste his time booming for days on end, out of season, when he could be high-tailing it to the feeding grounds. I didn't think that they are younger males practising for the important mating ritual in the breeding grounds down south. To me it seems less plausible that these are adult males currying sexual favours for the next winter holiday in the Caribbean than the possibility that it is the grandmothers fulfilling their role, singing to bring together females with calves needing her protection against the orcas up north.

Later in that season we came across an adult whale on the edge of the Bermuda platform. We guessed she had been feeding. She breached three times and then lay on her back and slapped both pectoral fins on the surface of the water. Almost immediately we saw a calf about half-a-mile away, in shallow water of 50 feet (15 m), lob-tail more than twenty times. Its mother rolled over and made her way over to the calf as the calf continued to lob-tail. Anticipating getting some underwater video of the mother and calf, I was in the water several times in their vicinity, with the engine off, expecting them to come to me. Often, I can get a curious whale to come to me, by diving and making sounds underwater. At times we drifted into very shallow water, as little as 30 feet (9 m) deep, but the calf never approached us although it was sometimes as close as 100 feet (30 m). We followed the mother and calf in the boat as they travelled to the platform edge where the mother breached several times again. Within minutes a large whale appeared at her side and they disappeared.

It seemed to me that the mother and calf we had just encountered breached, lob-tailed and slapped their pectoral fins as a means of communication to signal their location to one another. I had seen scarring on the calf's back. In the past I've noticed that calves with scars tend to be very shy and their mothers very protective.

Once again, it wasn't until we returned home that evening and I was sorting through the photographs that I noticed the calf had an entanglement around its head. In the photos you can readily see the scarred back of the calf, and closer up views of its head show a green polypropylene rope or net hanging off its back. If the rope was caught in its mouth it would have destroyed the delicate baleen used to filter the water. Although the calf was still probably able to obtain milk from its mother, this polypropylene net will not degrade. The calf will slowly deteriorate until it dies or is attacked by predators. The mother was emaciated, much more than other starving whales coming up north from the breeding grounds. I guessed she was spending too much energy fending off predators or staying near her injured calf. She had a large, visible chunk taken out of the trailing edge of her right fluke. This wound was relatively recent judging by the flesh still hanging from the fluke by a strip of skin. I presume the bite was made by a shark.

For the second time in a week it was a sad evening, once I recognized the calf's predicament. The debris we leave in the oceans kills these innocent and gentle animals. The calves with their innate curiosity and lack of experience are particularly prone to playing with seaweed, or buoys attached to ropes. No wonder this poor calf and its mother were in no mood to play and shunned our human presence. They had good reason.

That spring, on three occasions we thought we had lost whales that we had encountered (despite calm conditions and flat seas). On all three of these occasions we soon discovered the missing whales after checking the depth sounder. The first time this happened, we just happened to look at the depth sounder and watched as it changed from 170 feet (52 m) to 95 feet (30 m), 76 feet (23 m), 43 feet (13), 19 feet (6 m), 7 feet (2 m), and a fraction of a second later, a whale breached alongside us. With a 300mm lens on my camera (great for distant fluke shots, terrible for a close-up of a whale trying to get into the boat) all I can see was an eye and part of the curve of its mouth!

At the height of the migration, about mid-April, we found two large groups of humpbacks on Challenger Bank, swimming in very tight formation with some surface activity. One group consisted of 12–14 members and the other 14–16 members. The challenge counting this many whales together at the surface was eased somewhat because all of the whales within any one group tended to surface at the same time. Although by no means a definitive means of determining the sex of a humpback, judging by the well-formed dorsal fins on some, presumably the females, and the flat, battle-scarred dorsal fins on others, the males, one group appeared to consist of more than one large female, probably with some juveniles of both sexes and some adult males. In other words, this was not a group of males chasing one female as we often observe when we encounter large groups in the breeding grounds. They coordinated their surfacing every five minutes, blew several times and then dived almost simultaneously, showing their flukes. They did not head in any particular direction, careening all around the crown of Challenger Bank during the course of the day. This encounter provided a gold mine of twenty-eight individual whale fluke IDs, the most we had ever achieved in a day.

(page 114) Not only is Bermuda an important stop en route to the northern feeding grounds, it also appears that some whales, especially females either too young or too old to breed, may be staying in the waters around these mid-ocean seamounts during the winter.

Compared to the over-development on land, I'm always struck by how unchanged the ocean is, despite all that we have done to it over the centuries.

It seems that over the years the humpbacks assemble themselves into loose social units in the waters surrounding the mid-ocean seamounts before travelling together further north.

(top) This whale had a peculiar way of slapping its pectoral fins against its body as if it were injured.

Given that roughly 30 per cent of North Atlantic humpback whales have bite marks on their flukes, or their pectoral or dorsal fins inflicted when they were calves, there is clearly a compelling incentive for the whales to group themselves together in large, protective social units in the mid-ocean before heading north.

If the humpback whales are indeed gathering into loose protective social units around the mid-ocean seamounts before travelling north together, then their social ties are far stronger than has previously been supposed.

The next day we went out and spotted numerous whale blows but the whales were elusive. Nevertheless we obtained sixteen individual fluke IDs, eleven of which were the same whales from the day before.

In 2008, I had accompanied the US National Oceanic and Atmospheric Administration (NOAA) team satellite-tagging five humpback whales on Silver Bank, in the Dominican Republic. Unfortunately all those tags fell off the whales in a matter of weeks, before they left the breeding grounds. In 2009 the decision was made to tag whales later in the season so there would be a better chance of the tags staying on during the whales' migration north. The NOAA/Tracking of North Atlantic Humpbacks (TONAH) Project tagged another six humpback whales – five females and one male.

Dr Phil Clapham, NOAA's Program Leader for the Cetacean Assessment and Ecology Program in the US National Marine Mammal Laboratory (NMML), sent me regular emails with the updated satellite-relayed positions of these whales received from Dr Alex Zerbini, a National Research Council postdoctoral fellow working as a contractor with NMML's Cetacean Assessment and Ecology Program. It was a long shot, but they were hoping that if any of them came by Bermuda, I might be able to identify them and get a photo of their flukes to ID.

I sent an email to Dr Phil Clapham, describing my work and some of my thoughts about the humpbacks' pelagic migratory behaviour. His reply boosted my morale and sense of confidence:

This is all fascinating, and it's wonderful that someone local is now doing this work. I strongly suggest you write this up at some point. There are a couple of possible papers or notes in here, one of which would potentially look at the migratory destinations of Bermuda humpbacks to see whether they are biased (as one would expect) to some of the western North Atlantic feeding grounds. Speaks to migratory routes. I agree that not having preconceptions is a very good thing. Those of us who have been in the field forever confuse an excess of knowledge with common sense sometimes, or rather one can obscure the other. Fresh takes and perspectives are great, and as far as I know you're the only one who has been able to look at the social dynamics of humpbacks actually on a migratory route (I don't include E and W Australia here because the densities are very high and while that's a migratory route it's pretty clear that behaviorally it's indistinguishable from what goes on in a breeding ground). Cool stuff – keep it up!

At the same time as we had the two large groups careening around Challenger Bank, one of NOAA's whales, No. 87632, tagged on 6 April on Silver Bank, Dominican Republic, set off on her northerly migration only to slow down and remain for some days about 150 miles (240 km) west of Bermuda. Why was whale No. 87632, a mother with a calf, hanging around there? I wondered if she was feeding, or perhaps she was waiting for a convoy of whales before continuing her northward migration towards the home of the East Coast orcas. Maybe she was listening to the singer on Challenger Bank.

I then had several fourteen-hour back-to-back days on the water and I obtained more fluke IDs and more underwater footage of a calf and mother. When I returned home I had an email from Dr Clapham with an update of No. 87632's position. This mother and calf had changed course and were heading east towards Bermuda. Whether she had hovered over a seamount first, I could not determine.

When I emailed Dr Clapham to ask him what he thought No. 87632 was doing, he replied that as a scientist he didn't know, but his gut reaction was that she was feeding. But if she was feeding off a seamount like Challenger Bank, surely she would swim directly there, especially if she was with a newborn calf. It seemed strange that she would remain for some days to our west and then head off in almost the opposite direction to her migratory route, towards Bermuda. It is hard to imagine that she knew at that distance there was food available in Bermuda. She then stayed to our immediate north-east before continuing northwards towards Newfoundland. Once she hit the continental

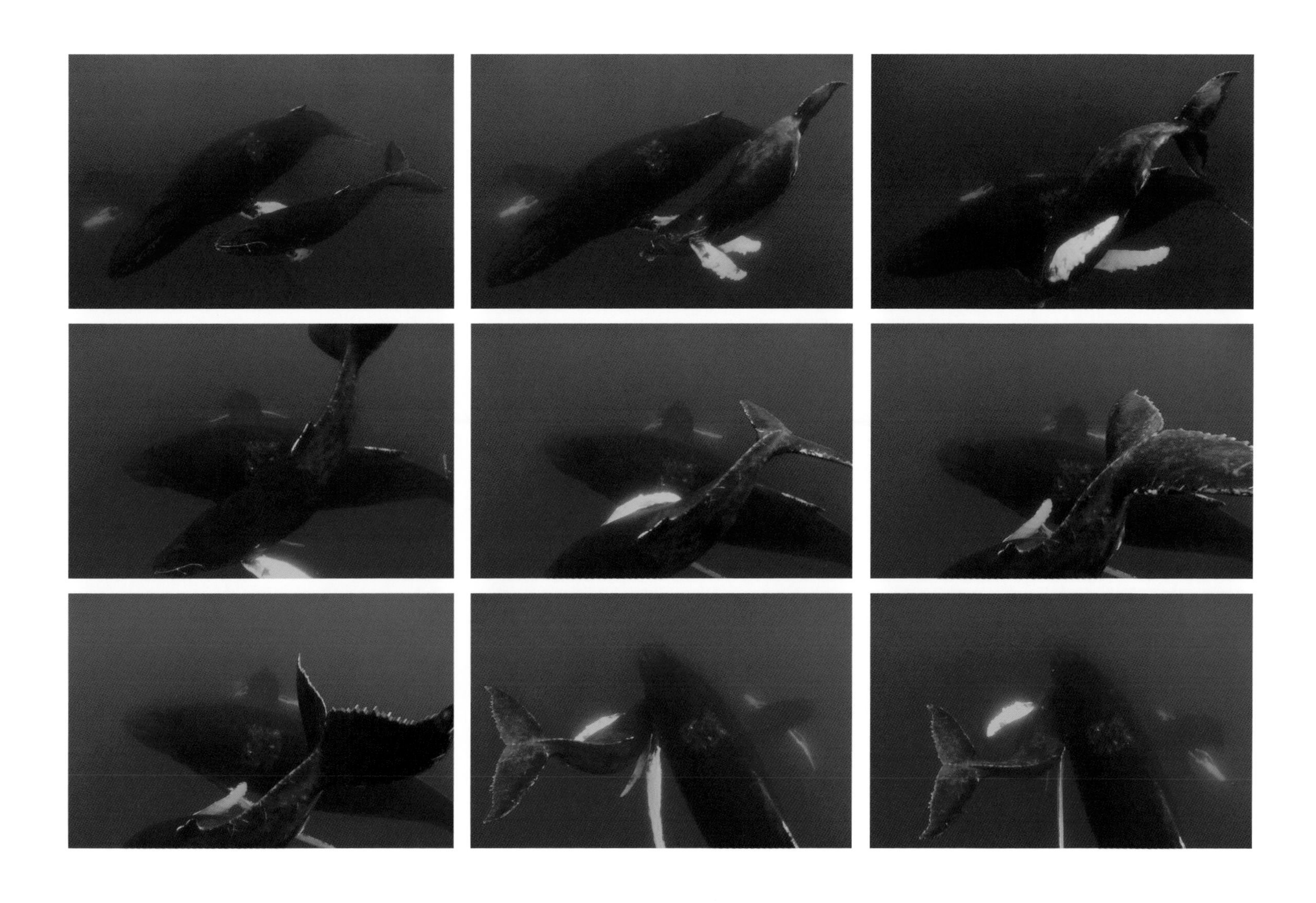

We often see mothers and calves seemingly waiting. Is this in order to form part of larger groups before running the gauntlet of orcas further north?

Whenever we saw a mother and calf without an escort, it was not long before an escort appeared.

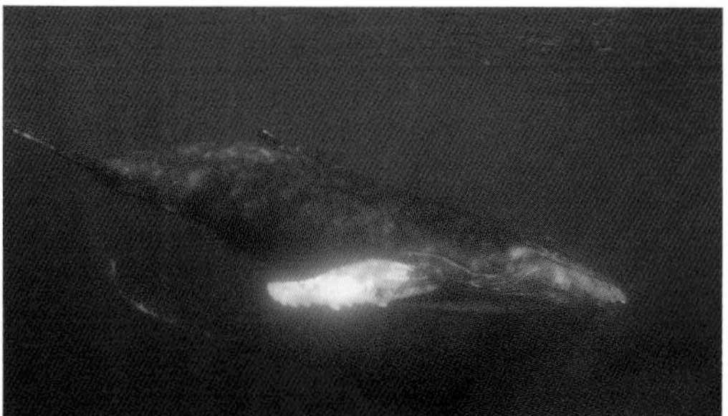

shelf south of Newfoundland, she turned 90 degrees west and crossed the Laurentian Trough towards the Scotian Shelf.

My guess was that she had heard the singing on the crown of Challenger Bank, and then headed towards Bermuda and remained just north-east to intercept the singer as it continued its own migration northwards.

Later that season there was a two-day window of calm weather, so we set off early one morning on *Sea Slipper* for another overnight trip. We saw several bait balls (pilchard-sized fish huddled together for protection against predators) on the surface, within a hundred yards of some humpback whales exhibiting behaviour commensurate with feeding.

We dropped the hydrophone overboard several times that day and made one particularly beautiful 25-minute recording of a whale singing. We moored around 6 p.m. in 170 feet (52 m) of water on the Challenger Bank seamount, 15 miles (24 km) off Bermuda, and began rolling watches, three hours on, three hours off. Throughout the night we dropped the hydrophone overboard and recorded whale song. At around 4 a.m. it was dead calm, the moon had gone down and the stars were bright; Michael and I were on watch listening to the gentle lapping sounds of the waves against the hull of the boat when we heard the massive blows of whales breathing right beside the boat – a magical, mystical sound.

In the morning we spotted a whale in the distance, drove over to where we had seen it, put the engine in idle and waited. I vaguely imagined I heard a whale singing but thought it was just wishful thinking. I recounted to the others on the boat how Jim Darling in Hawaii had parked his boat right on top of a singer and how we could hear the whale singing through the hull of the boat. We sat there quietly looking for the elusive whale when we realized we could hear the whale somewhere underneath us. Despite being on a big boat on the upper deck, with the engine idling, we heard a whale singing through the hull. Michael thought I had played a trick on them and was playing back the recorder with songs we had heard during the previous night. We turned the engine off and eventually the whale surfaced some 90 feet (27 m) away. It was the only song we heard that day, but was impressive for being heard so readily through the hull of a trawler.

The window of good weather continued and I spent eight out of ten days on the water, some days fourteen hours long, in addition to the overnight trip on Challenger Bank. This was one of the many highs this season. The lows were the long hours of looking before we found anything. Twice, baby whales found us. In both cases the humpback calves disappeared mysteriously, without a trace, on perfectly calm days. I reckoned they were using re-breathers, buddy-breathing or frequenting an oxygen bar. At least, that was my hypothesis.

And that was the end of our third and final year filming for the Humpback Whale Film Project and the last time we'd have the opportunity to film humpback whales before I put the footage together.

Sadly, the last whale that was still transmitting its position via satellite stopped doing so. All the satellite tags were now dead. Some of the information the tags beamed back to us, especially the one on No. 87632, was unexpected. But just when we thought the season was all over, another whale started transmitting for the first time, west of the Azores. Did she come by Bermuda? We will never know. She continued transmitting until it looked as if she was bypassing Iceland to the east and heading for Norwegian waters. A surprise for all of us, especially for those who didn't think humpback whales in the Caribbean fed in Norwegian waters.

An adult male cuts me off from the curious calf and flicks his tail at me as he swims by as if warning me not to follow. This escort seems so close to me from the surface and yet looking through the viewfinder of the Gates housing underwater, it appears relatively small and far away.

CHAPTER NINE

CANDLE AND HARRY POTTER REAPPEAR

Around mid-March 2010, I had a phone call that a whale had washed up on the rocks in Somerset. I phoned Camilla Stringer, a loyal volunteer, and we drove to Somerset to climb along the rocky shoreline looking for the whale. Eventually we found its body being battered against the rocks. It was a dead baby sperm whale. There were no obvious markings on it to account for its death although there were plenty of plate-sized shark bites taken out of it, probably after death. After calling up the vet at the Aquarium, I tied a rope around its tail and we towed it slowly behind a fisherman's boat to Ely's Harbour for a necropsy to see if we could find out why it had died. Because inclement weather was expected, no necropsy was performed and the carcass was towed back out to sea, unexamined, a day later.

I have now seen a dead humpback, a dead sperm whale, a dead pygmy sperm whale and a dead bottlenose dolphin on our shores, and Bermuda is just a pinprick in the ocean. I wonder how many more animals die out there, never washing ashore, and why they die.

Because of consistent bad weather and because I had been sitting in front of computers making the final edits on my movie, *Where the Whales Sing*, we set out on *Sea Slipper* for the first time of the year much later than usual. Sadly, the first whale we encountered, in 50 feet (15 m) of water barely 2 miles off Somerset, was an entangled whale.

We saw this humpback breach about fifty times, but each time it breached it only exited the water to its dorsal fin and then flopped onto its chin. It did not pirouette onto its back and whack the water with its rotating outer pectoral fin as most breaching humpbacks do. Occasionally we caught sight of something on one side of the dorsal fin and then on the other side and we could see the whale's back was scarred. We hoped to get a fluke ID photograph, but our failure to do so was a clue that all was not right. Despite multiple breaches, eventually in deep water, the whale never showed its flukes. Something was wrong.

We were idling when the whale breached right beside the boat and we could see that the pectoral fins were pinned to the sides of the animal and that they were pink from abrasions. The humpback was very entangled. A quick examination of the photographs on my digital camera revealed that the whale was caught in black, polypropylene rope. On one side there was a red buoy attached. The rope seemed to be a single line that had gone into the whale's mouth, like a bit on a bridle in a horse's mouth. The rope twisted around the whale's body, pinning its pectoral fins against its body while clumps of rope draped in tangled knots over the whale's back and tail. Whales use their pectoral fins to angle their bodies up and down when surfacing and diving so it was no wonder this whale didn't curve its back the normal way, showing its flukes as it dived. The photos also revealed its partially opened mouth as it breached, an indication that the rope was exerting a considerable force on the lower jaw to force it open like that. The pain this whale must have endured is unfathomable.

When I examine my photographs carefully, most of the whales I have photographed seem to have entanglement scars on their backs, dorsal fins and peduncle or tail area. The sight of this entangled whale had me up most of the night as I imagined the torture the whale had to endure with its pectoral fins pinned painfully to its sides and a rope tugging at the corners of its mouth. It was an ominous sign when the first whale we encountered and photographed of the season was badly entangled – the third entangled whale we had encountered in Bermuda over three years.

It was, once again, a graphic demonstration of how our oceans are negatively affected by human activities; not only littered with lethal ropes and nets, but also filled with noise and chemical pollution, poisoned by toxic plastics, altering their natural state through climate change and ocean acidification. These negative impacts on marine environments must inflict a kind of collective psychological anxiety upon whales, much as they do in our own pressured societies.

There is something special about whales, which makes us to want to help and to protect them. Perhaps it is sympathy for creatures that have been relentlessly driven to the point of extinction, or perhaps it is it our awe of their size and their intelligence. Whatever it is that pushes us towards the whales, they remain an enigma, their underwater lives still secret and unfathomable.

In 2005 there was a front page story in the *San Francisco Chronicle* about a female humpback whale that had become entangled. Weighed down by hundreds of pounds of crab traps and lines wrapped around her body, her tail, her torso and a line tugging in her mouth, she was struggling to keep her blowhole above water. A fisherman spotted her just east of the Farallon Islands and radioed an environmental group for help. Within a few hours, the rescue team arrived and determined that she was so badly entangled that the only way to save her was to dive in and untangle her. They worked for hours in the water to cut her free. Freed of the ropes, the divers described the whale swimming to each and every diver, one at a time, to nudge them gently as if thanking them. The divers said it was the most incredibly beautiful experience of their lives.

Recently I had dinner with a visiting marine biologist in Bermuda who described to me how scientist colleagues of his in the Antarctic had witnessed a seal being chased by orcas. A humpback entered the scene, appeared to protect the seal from the orcas, and eventually rolled onto its back so the seal could climb onto its stomach, safely out of reach of the orcas. There were photos of the event as it unfolded and a scientific paper will be forthcoming. I don't have any doubt about this story, or that a humpback would have sympathy for a seal being pursued by orcas and come to its rescue.

After weeks of bad weather, just when we thought this season was going to be a bust, we managed to get out on the water every day for a week. Most days we saw somewhere between twenty to fifty whales, with anywhere from fifteen to twenty individual fluke IDs obtained as well as underwater footage. Getting out on the water five days in a row was invaluable for the all-important intra-year re-sightings. We also matched some of these fluke IDs to previous sightings elsewhere, and some to re-sightings here in Bermuda, both from year-to-year as well as within the same season. In the three years that we have applied ourselves to taking fluke identification photos we had our third home run – a whale that had been photographed in Bermuda three times in the last four years. If we were able to get out on the water every day during the peak six-week migration period, I am sure we would have had many more re-sightings.

With all of our re-sightings from year to year there is remarkable similarity in the dates from one year to the next. Unlike us, the whales don't use Gregorian calendars and are likely migrating in time with the phases of the moon. Given that we also know the whales are here for some days and that the moon phases are not synchronized to the Gregorian calendar from year to year, it is remarkable how the re-sighting dates closely match from one year to the next. It would seem some whales not only maintain fidelity to their migration routes but also adhere to a specific timetable. Both observations support the idea that the whales are meeting on these mid-ocean seamounts to gather into known social units to continue north. In other words they don't just appear haphazardly on the same feeding grounds year after year; they gather in the middle of the ocean and travel to their summer feeding grounds together, possibly as a defence against orca attacks.

Generally it seemed we might have two 'waves' of whales per moon cycle during the height of the migration. These waves seem consistently to coincide with the full and half-moon. As I mentioned, we can see scores of whales one day and the next day see almost none. If the whales do aggregate here into familiar associations before moving on, it would make sense that they time their departure from the breeding grounds at the same phase of the moon, year after year. Looking at the dates that the whales were re-sighted here in Bermuda, there are remarkable patterns – most re-sightings are within ten days of the date of previous re-sightings. The re-sightings out of that window are the exception to the rule.

I continued to look out for Magical Whale, but our mysterious friend had simply disappeared.

My two most stalwart crew members, Michael Smith, the owner and captain of *Sea Slipper*, and Camilla Stringer, who had helped me from the first year and joined me on most of my trips looking for whales, were so enthusiastic that on occasion they were keener to get out on the water than I was. The only reason for my hesitation in heading out in any kind of weather was the extra burden I placed on Annabel to cover for me, having to pick up our two daughters and take care of them after she had already endured a hard day at her medical practice. By the time I arrived home, especially if I were out on *Sea Slipper*, both children were usually asleep. I didn't want to place this extra workload on Annabel unless I felt that we would have a productive day on the water.

One morning Michael phoned to suggest we go out. It was a grey and very windy day and despite my better judgement Michael persuaded me to pack my gear and pick Camilla up. As soon as we set out on the open ocean I regretted the decision. We began rolling heavily in the ocean waves as we trolled along the edge of the Bermuda platform. There was very little chance of coming across any whales. We'd never see their backs amongst the giant waves, or their blows with the high winds and the grey, dull light. After some hours of fruitless searching we were heading toward Challenger, heaving from side to side at alarming angles. I am blessed with a cast-iron stomach. I don't get seasick, and we warn any volunteers who come out with us that we don't head back if they are seasick. We have proven good to our word on many occasions. I enjoy being out on the water most of the time, and windy, wavy conditions are just fine, but this seemed like a total waste of time. As I stood behind Michael, holding on to anything to keep my footing, a whale breached right in front of us. Michael punched the GPS, marking our position. Although we knew there was a whale around us, we lost it. We did circles and continued searching vainly for some sign of the whale. Eventually I sat dejected at the back of the boat and was staring aft when a whale breached in our wake, as if to wake us up. Then he started lob-tailing and as soon as he did that I recognized him. I took photo IDs of his flukes and when he eventually disappeared I went below decks to check the digital image in my camera against my laptop containing all of our fluke IDs. I knew this whale and the photos proved it.

It was Candle, our No. 0002, the one that was with Magical Whale and me in 2007. No. 0002 or NAHWC No. 1135 was first photographed by Dr Hal Whitehead in Newfoundland in 1978 and not seen again until we photographed him on 23 April 2007. Shown in my film, *Where the Whales Sing*, this is the whale that approached me when I was in the water with Magical Whale. It was a profound experience to see this whale again and I immediately felt an intimate connection with him, knowing I had been swimming with him four years earlier. This re-sighting was within a calendar week of our previous sighting of Candle in 2007.

While I was selecting the photos for this book I noticed that I had photographed another whale, No. 0004 that same rough day on the water in 2010. By coincidence, I realized that Kelly Winfield had photographed No. 0004 on the same day and roughly about the same time that we had photographed Candle and Magical Whale in 2007. We had evidence now of the same pair of whales, No. 0002 and No. 0004 in the same location and at the same time *(previous page)* A whale spy-hops beside a spouting whale.

(this page) A female will breach to encourage its calf to do the same thing, possibly to help it build up its strength for the long journey north to the feeding grounds.

Having spent so many hours in the water trying to get underwater footage for my film, I have witnessed many unexpected events. We pulled this dead baby sperm whale in for a necropsy. On another day we found a chunk of blubber floating in the ocean and wondered where it had come from.

travelling together four years apart. Candle is at least thirty-two years old and probably a male like Magical Whale, and the other whale wasn't a calf. It can't be just a coincidence that the two were together in the middle of the ocean in both years. It was yet another piece of the jigsaw puzzle and once again seemed to confirm the likelihood that the whales do use the seamounts to aggregate into socially cohesive groups.

Whale No. 0004 is interesting for other reasons. It's another of our home runs, photographed in 2007, 2009 and 2010. Considering I obtained only fifteen fluke IDs in 2007, these re-sightings from that year are even more remarkable. We also have intra-year re-sightings: in 2010 No. 0004 was photographed on Challenger Bank on 8 April and 13 April, six days apart. Obviously it was in no rush to go anywhere.

Was Magical Whale present with Candle that rough weather day in 2010? Perhaps it was Magical Whale and not Candle that had breached first, in front of the bow. It was the same location I had swum with Magical Whale and Candle four years ago. With two of his mates from 2007 back again four years later, I'd guess Magical Whale was probably there too. I'd like to think so. In 2010 I had little doubt that Candle was communicating to us. In an extensive, windswept ocean he (or perhaps Magical Whale) had chosen to breach right in front of our bow. When we couldn't find him, Candle breached in our wake, 50 feet (15 m) behind us, and then began lob-tailing for several minutes as if inviting me to get in the water. I believe humpbacks can identify different boats by their sounds, but I was on *Sea Slipper* and not *Pheidippides*, as I was in 2007. Do I believe Candle actually knew I was on the boat? That's a difficult question to answer. I don't rule out anything. The more pertinent question by a non-sceptic might be – how did he know? Seeing Candle again was very special and I didn't mind not seeing Magical Whale. In a way I want him to remain an enigma.

Whales don't have opposing thumbs and forefingers to build cell phones, satellites and computers. Although the humpbacks do not have the same brain size to body mass ratio as us, they do have big brains, much larger than ours with all the spindles, neurons and cortex components put together in as complex a structure as a human's brain. Being big doesn't guarantee a big brain. Dinosaurs were huge animals but they didn't have huge brains. The whales developed their complex brains for a reason. They must use all that processing power for something. Their intelligence manifests itself in other dimensions that we have yet to understand. Imagine the brain power involved in echolocation whereby a dolphin can send out a signal, receive the echo and process the loudness, frequency and timing of that echo to identify a golf ball sinking in the water more than a football field away, while moving directly towards it. Just as we have difficulty comprehending humpbacks' intelligence, I suspect the whales and dolphins have equal difficulty grasping manifestations of our human cleverness. Humans' inability to live in harmony with our environment might lead the whales to think that we aren't smart at all.

The skeleton of a humpback's pectoral fin shows remarkable similarities to our own hands and fingers. The humpbacks have the longest fins of all whales, a third of their body length. They use them in ways comparable to how we use our own arms and hands. They use their 16-foot (5 m) pectoral fins as weapons, just as we might use our arms in martial arts. Humpbacks use their fins to cradle their calves, just as we use our arms to hold a child. When I swam with Magical Whale I could see the tips of his fins flex delicately when he was reaching to me, just as we can bend our fingers to grasp something. Perhaps we have this greater connection with humpbacks because we are hard-wired to use our hands and fingers, similar to the humpbacks, although they use arms and hands that have been modified into pectoral fins.

Before the season was entirely complete, we had one more uplifting encounter that balanced the awful sight of the first humpback we saw this season – the entangled whale. It was mid-May and we had seen whales on every outing since we started our field research in late March. We had accumulated another 120 individual whale fluke IDs despite not getting out on the water as often as previous years. I suspect this is because we were getting better at finding and photographing the whales, and also because the number of whales off Bermuda is increasing. Just as we left Somerset a land-based spotter directed us to three whales in shallow water off South Shore. The three whales came to the boat several times, which allowed me to film them underwater. It looked like a mother and calf, and escort. As I free-dove down to the whales the escort rolled 360 degrees onto its side and then onto its back as it came towards me, thereby providing a perfect view of its unusual belly pigmentation, which resembled a white chevron. When I studied the underwater video that evening it struck me that the white pigmentation wasn't genetic, but rather the scarring from an entanglement. Looking closely through the video I could see white lines descending from both sides of its mouth to the underside of its pectoral fins and across the belly in a wide curvilinear scar.

Being the eternal optimist, these final images from this season left me with a sense of hope. The first whale we saw this year was badly entangled with a rope through its mouth, pinning its pectoral fins to its side and a bundle of rope swaying from one side of its back to the other with more trailing off its flukes. The male escort we filmed today had scarring that must have been from a similar entanglement that was just as bad, including a bridle through its mouth. Perhaps that unfortunate whale we encountered early this season had been as lucky as this one. It might have been released by human rescuers, or it might be one tough survivor.

When I returned home I sent a photo of this whale's dorsal fin to Dr Jooke Robbins, Director, Provincetown Center for Coastal Studies, and, as she often does, she responded within minutes with the ID. The result was confirmed by the fluke ID photograph I sent her later. It was Pendiente, or NAHWC No. 0523, first photographed in Maine in 1987.

Some days later Judie Clee made another remarkable match. A photo we had taken on 12 April 2010 matched within our own catalogue to No. 0039, a whale we called Harry Potter because of the lightning mark on its fluke. This is a whale we first photographed here in Bermuda in 2006, then again in 2008 and 2009, and now in 2010. Three of those sightings were within the same two-week period as measured on the Gregorian calendar. Harry Potter is matched to Allied Whale's No. 1517, first identified in 1979, thirty-one years ago in Newfoundland. Harry Potter's sighting history from the North Atlantic Humpback Whale Catalogue is as follow:

25 Jun 1979	**Newfoundland (Memorial University of Newfoundland)**
13 Aug 1982	**Labrador (Memorial University of Newfoundland)**
12 Feb 1983	**Silver Bank (Ken Balcomb, ORES-Ocean Research and Education Society)**
13 Aug 1983	**Labrador (Memorial University of Newfoundland)**
29 Jul 1991	**Witless Bay, Newfoundland (Memorial University of Newfoundland)**
Jul 2001	**Witless Bay, Newfoundland (Wildland Tours)**
2 April 2006	**Bermuda (Mark Outerbridge)**
11 April 2008	**Challenger Banks, Bermuda (Andrew Stevenson)**
29 March 2009	**Bermuda (Lindsay Smith)**
12 April 2010	**Challenger Banks, Bermuda (Andrew Stevenson)**

Later in the summer of 2010, Reg Kempen, a retired British banker and whale enthusiast in Newfoundland, emailed me, as he usually does with all his fluke IDs. This time he sent a photograph of a fluke taken by Kris Prince, a young commercial whale-watch operator, off Spaniards Cove (just North of Bonaventure Head), Trinity Bay, on 9 July 2010. As soon as I saw it I recognized Harry Potter.

(previous page) A whale breaches off Bermuda. We have seen Harry Potter (named for the lightning scar on the whale's fluke) five out of the last six years.

We have seen Candle, the whale which was with Magical Whale, two years in a row since 2007.

This whale's pectoral fins are tied to its sides and knots of rope trail from the bridle in its mouth. Although the whale breached some fifty times, the rope remained embedded in its mouth. Although these whales are tough and do survive such entanglements, the torture they must endure defies the imagination.

As I free-dived down to the whales, the escort rolled 360 degrees onto its side and then onto its back as it came towards me, providing a perfect view of its unusual belly pigmentation, which resembled a white chevron. The white pigmentation wasn't genetic, but rather the scarring from an entanglement similar to that shown on the opposite page.

(left) By the end of the 2011 season, over a period of five years, we had identified more than 500 individual humpback whales by the fluke patterns on the ventral sides of their flukes. Twenty per cent of these we matched to fluke IDs we had taken of the same whales over the previous four years. In 2011 we photographed two whales over a period of eight days, several over six days and many more over shorter time periods. Clearly the humpbacks are in no rush to by-pass Bermuda.

(below) I discovered during the 2011 season why the whales were disappearing in shallow water – they were having sand baths in sand holes, scraping the dead skin and sea lice off their bodies.

CHAPTER TEN

WHAT DOES THE FUTURE HOLD FOR MAGICAL WHALE?

In the summer of 2010, I took my family to Bar Harbor, Maine, an area I had become fond of since first coming to visit researchers at Allied Whale, College of the Atlantic in 2007. Elsa took summer courses at the College, leaving Annabel and me to hike around the Acadia Mountains with Somers, not yet two years old. We also cycled, while towing Somers behind us, on the numerous carriageways reserved for bicycles. From Bar Harbor we drove to New Brunswick, Canada, to stay at a friend's house in St Andrews at the mouth of the Bay of Fundy where the tide ranges almost 40 feet (12 m), the highest tides in the world. We spent most of our time on the wharf, sometimes as much as twelve hours a day, collecting crabs, shells, starfish, sea ravens, sea urchins and fish for Elsa's impromptu touch tank/bucket. She even managed to catch a lobster. Elsa, along with her ardent admirer and sidekick Somers, charged a nickel for adults to look into her 'touch tank', but kids were free. She was in her element explaining the fish from information she had picked up from passing fishermen and other locals. We became fixtures on the dock, especially at low tide when the pickings were good. I think another week and Elsa would become the wharfinger, the keeper of the wharf. We also went looking for whales and found numerous fin whales feeding. From St Andrews we took the ferry across the Bay of Fundy to Digby and then down the spit of land that ends at Brier Island.

Elsa, now seven, was much more interested in whale watching by this stage, especially when we went out in a small Zodiac, and she participated in photographing and videoing the humpbacks that often came right up to us. At one point Elsa abruptly threw her small Nikon point-and-shoot camera on the bottom of the Zodiac and yelled in frustration, 'I hate this camera! It takes too long to take the picture! Please let me use yours, daddy.' I did let her use my Nikon D200, but not all the time, much to her disappointment.

As had happened before, we had multiple great experiences with the whales in Brier Island thanks to the hospitality of local residents. Elsa loved being able to cycle along the main street without having to worry about traffic. It was all in all an idyllic holiday blessed with clear blue skies. On our last trip out looking for whales, we came back to Westport Harbor to find a female right whale and calf in the harbour. Hunted to the point of extinction, there are only some 400 of these northern right whales left in the world. Ship strikes and entanglements keep whittling the numbers down despite rerouting of shipping lanes so that traffic avoids areas often frequented by the right whales. There is some question as to whether the North Atlantic right whale population will ever recover. Elsa and I were ecstatic to see these whales.

By now I could answer Elsa's first question: Why does a whale breach?

When children ask me this question, I like to respond that I'm not a whale, so I'm only guessing. There is no doubt in my mind that breaching is sometimes used as a means of short-distance communication, to let other whales know where they are. Breaching is probably also used to dislodge patches of dry,

(Top left) Elsa holds a section of humpback baleen.

(Middle row left) The next generation will know more about whales at an earlier age.

(Middle row right) Here Elsa and Somers sit at interactive displays at the Bar Harbor Whale Museum.

(Bottom two rows) The good news is that the increasing number of whale-watching operators will foster goodwill towards the whales. The bad news is the growing number of entanglements all whales experience as we attempt to fish ever further from the shore.

The lines attaching tens of thousands of lobster pots to surface buoys are a major threat to the foraging whales.

Elsa smiles after a successful whale-watching trip. This enthusiasm will, I hope, lead her to help protect the whales' right to live peacefully in the ocean.

sunburned skin which you can often see on the surface after a whale breaches. Or perhaps they breach to get rid of sea-lice which I have also seen lying on the surface of the water where a whale has breached. I've seen whales breach to get rid of the dolphins riding on their noses. They also seem to breach to show off to females or to intimidate other males. I have seen them breach to try and disengage an entanglement. But the reason I like to give to kids is that the humpbacks, especially the younger ones, breach because they can, just for the sheer fun of it.

On our last afternoon Elsa had her own Magical Whale encounter. Annabel, Elsa and I were in a Zodiac. We saw a mother and calf and turned the engine off as they approached. It was the beginning of a remarkable encounter where the calf serenaded Elsa just as Magical Whale had serenaded me four years earlier. The calf came to the Zodiac and swam around and under it, spy-hopped us, rolled on to its side and, with its one eye out of the water, slapped the water as if asking Elsa to do the same. It played with seaweed balanced on the end of its nose and came to the boat as if offering it to Elsa. At one point the calf lifted its fluke out of the water and rested it against the Zodiac right beside Elsa. Were it not covered in barnacles she could have patted it. Then it breached right in front of us and then almost immediately afterwards did a tail breach. It seemed that this calf knew there was a small child in the Zodiac and related to her because of that. The calf played around us for more than an hour while its indulgent mother whom we later identified as Shuttle, first photographed in 1986, stayed to the side.

This spirited young whale played around us, apparently trying to communicate with Elsa, despite the ropes and debris we humans leave in the ocean to traumatize these friendly creatures. The magical interaction between the calf and Elsa was an uplifting experience but it also left me with an overwhelming sense of sadness at the suffering I know we inflict on sentient, intelligent animals such as this calf and its mother. Humpbacks can probably live almost as long as humans. Would Shuttle's calf survive all the hazards we put in its way to live as long as Elsa?

Immediately after that encounter we returned to Bermuda where I departed a couple of days later to attend the BLUE Ocean Film Festival. Having just had a month's holiday I didn't feel I should take off again, but emails from the festival director asking me to attend, with the convincing promise that I wouldn't regret it, had me on the next flight.

Six months earlier, when I completed *Where the Whales Sing*, I was exhausted. I'd spent four to five months sitting in front of computers for more than ten hours a day, teaching myself how to use the editing software so that I could edit my film. Elsa's narration of the film at six years old was touch and go at times. I didn't want to force her to do it, but she was determined to stick with it. 'If someone else did it, it wouldn't be true,' she told me, quite rightly. Despite having a bad head cold and two missing front teeth, she sat in the clothes closet and I recorded as she read the script from my netbook. When I finally finished editing the film I told Annabel that I'd never make another film again. I should know better. Taking the seminars offered at BLUE (given by some of the best-known and most skilled producers, directors, filmmakers and cameramen) re-inspired me. It was an invigorating week spent amongst some of the best underwater filmmakers in the world. By the end of it I was motivated to make another film. I phoned Annabel to tell her.

'I thought you weren't ever going to make another film again,' she replied.

'I've changed my mind,' I said.

'That didn't take long,' she commented, without surprise.

I took two mornings off to go whale watching in Monterey Bay. I saw my first blue whales without Elsa, much to her chagrin. There were also scores of humpbacks lunge-feeding on the krill. We were so close that I could see the krill jumping out of the water just before the whale's open mouth closed on them. I had never photographed this behaviour before and the only camera I had brought with me on this trip was Elsa's point-and-shoot digital Nikon. I repeatedly missed the best shots I'd ever have of a whale lunge-feeding because

Elsa's photographs of the friendly calf breaching and tail-lobbing beside our Zodiac.

Elsa had her own Magical Whale moment when this calf frolicked for hours beside our boat. At one point it flopped its fluke against the side of the zodiac within reach of Elsa.

While the calf continued to play around our Zodiac, the mother watched patiently. Later we identified the mother by her fluke ID, finding a match for it in the North Atlantic Humpback Whale Catalogue which contains some 7,000 individual fluke IDs of North Atlantic humpbacks.

Outside of the whaling season there is still plenty to do. Elsa swims with the fish in shallow waters *(top right)*.

From the last week of December, we start looking for whales from the South Shore *(bottom left)*.

Somers, at two years old, can already distinguish between a distant whale and waves crashing on a coral head. If the weather is really bad, we can always find a pond in a garden somewhere to catch tadpoles and guppies *(bottom right)*, purely on a catch-and-release basis!

It gives me a great sense of well-
being to know that some of the
young calves I film or photograph
and identify, will be swimming in
the ocean long after
I am no longer on this earth.

of the long time lapse between pressing the button and the camera actually taking a photo. I understood first-hand Elsa's frustration at Brier Island. 'I won't do that to you again, Elsa,' I promised aloud, laughing at the poetic justice of the situation.

Of 400 films entered into the BLUE Ocean Film Festival, there were 19 award winners including films made by Disneynature, the BBC Natural History Unit and National Geographic. Elsa and I won the 'Best Emerging Underwater Filmmaker' category, which eliminated all the professionals but allowed us to be on the same stage. Gates, the makers of my underwater housing, were so impressed with Elsa's narration of the film that they gave her an underwater housing for a small HD video camera. She'd have no excuse to complain about her camera in the future.

At a recent screening of *Where the Whales Sing* she was asked if she still wanted to become a zookeeper. Her reply surprised me. 'I want to be an explorer and maybe I'll study the whales, like dad.'

Magical Whale has not been photo identified before or since my encounter with him. I reckon it's because he feeds somewhere where there are few photographers, like Greenland. During the summer of 2010, the International Whaling Commission (IWC) met and allocated to the indigenous peoples in Greenland a quota to kill twenty-seven humpback whales – nine per year for three years, beginning in September 2010. These are the first humpbacks to be killed in Greenland under IWC auspices since the moratorium in 1986. They are also the same humpbacks that migrate past Bermuda every spring. Magical Whale's attempt at interspecies communication that made him so popular here in Bermuda is the same friendly behaviour that will make it easy for someone in Greenland to skewer him at the end of a modern, grenade-tipped, explosive harpoon. By the end of August 2010 – even before the season officially began – the first humpback was killed for its meat by an eager crab fisherman.

I have made contact with schools and communities in Greenland and, if requested, send them DVDs of our film *Where the Whales Sing*. Whether or not humpbacks are no longer endangered, I want them to know how much we care about Magical Whale and all the humpbacks. Elsa has made pen-pal friends there. In the summer of 2011 I would like to go to Greenland with Elsa to look for Magical Whale.

Magical Whale has become well-known in Bermuda through Elsa's and my film. The film has been screened on local television over 500 times. DVDs of the film have been distributed to every school. Two thousand DVDs have been sold or otherwise disseminated in a population of 65,000 people. I have written numerous articles about humpback whales. As a result of this publicity Magical Whale has entered the hearts of many Bermudians.

I have given countless personal screenings of *Where the Whales Sing* to schools. I sit with my back to the screen and watch the children's open-mouthed reactions. If Magical Whale is killed in Greenland many young children here in Bermuda will be devastated at the loss of a friend – a mythical creature that has fired their imaginations and enhanced their awareness of our marine environment.

In the meantime, each winter and spring, I persist in looking for Magical Whale. One day I hope that Elsa will be with me when I find him. Perhaps she will be able to swim with Magical Whale, too.

That would be nice.

Elsa's simple question when we saw a whale breach off our beach six years ago triggered our passion for humpback whales. I have found some valid answers to Elsa's question as to why the whales breach, but perhaps it is Elsa's answer that is the most poignant – they breach because they can, just for the sheer fun of it.

The research group Allied Whale at College of the Atlantic has a catalogue of some 7,000 individual North Atlantic humpback whale fluke identification photographs. When a new fluke ID is sent in, someone takes that new fluke ID through the catalogue to either match it with an existing photograph, or assign it a unique identifying number as a newly identified whale. There have to be at least three points of matching identification. If it is matched, two of the Allied Whale staff have to confirm the match.

To try to make this process somewhat easier, there are five basic types of whale flukes:

1. **Mostly white.**
2. **25 per cent white.**
3. **Half white, half black.**
4. **25 per cent black.**
5. **Mostly black.**

With some luck, one might only have to look into one type – this would usually only occur in mostly white or black flukes. Often one can eliminate three types, and look through only two types. Sometimes one has to look through three types, especially if the whole fluke is not visible.

Allied Whale has come up with some sub-types that can make the process of identification easier. These include: typical; 'eyes' white surrounded by black; wide black trailing edge; wide black leading edge; 'fireworks'; white on trailing edge; straight-sided; rounded not to notch.

None of this makes much sense until you see the diagram on the right depicting these types and sub-types. Photographs of whale flukes are not always ideal. Sometimes they are dark, sometimes incomplete, often at an angle and not from directly behind, and sometimes the angle of the fluke is not vertical. Often a splash or reflected sunlight will obscure part of the fluke. There is no computer software program yet that can reliably identify whale flukes. It is still down to the human eye to make the judgement call.

In 2007 we spent about 30 days and some 200-plus hours looking for whales, mostly along the South Shore. We spotted whales one out of three days and obtained 15 individual fluke IDs. In 2008 we spent more of our time heading out by South West Breaker to Sally Tuckers and Challenger Bank. We spent 23 days on the water, over 200-plus hours, and estimated that we spotted 339 different whales of which we obtained 62 individual fluke IDs. In 2009 we again spent more of our time heading out from shore directly to Sally Tuckers and then across the canyon to Challenger Bank. We spent 20 days on the water, over 200-plus hours, and estimated that we saw 311 whales from which, with a new longer lens, we managed to obtain 168 fluke IDs. Because of bad weather, in 2010 we spent fewer days on the water, but still obtained 120 fluke IDs. In 2011 we spent 23 days on the water and obtained over 160 individual fluke IDs of which some 30, or 10%, of them were re-sightings to the previous four years. Once again, the re-sighted whales that migrated past Bermuda during 2011 returned within a week of the same date of their previous sightings. Our 450-plus individual fluke IDs in five seasons is more than triple the 146 Bermuda fluke IDs taken by visiting scientists and local residents in the forty years before we started our project in 2007. 45 of these individual whale IDs were re-sightings of two or more years.

We re-sighted whales often in the same season. The longest spread of a re-sighting of an individual whale here within the same season is eight days. In fact, two whales during the 2011 season were re-sighted over an eight-day period while many other whales have been re-sighted over six, five, four, three and two days. Looking at the re-sightings from year to year, we have one whale we have re-sighted over four years, three over three years and over a dozen over two years. Often the re-sighting date from one year to the next is within a week or so of a previous year's sighting. Of the 2007, 2008, 2009, 2010 and 2011 whale flukes, we have some 110 matched to the NAHWC catalogue and another 50 have been assigned a new NAHWC number as never having been seen before. This means at least half 'our' whales have never been identified before. Of the ones that have been identified and matched, their feeding grounds range from the eastern seaboard of North America – from North Carolina up to the Gulf of Maine, the Bay of Fundy, Nova Scotia, Newfoundland and Labrador – to Greenland, with two whales feeding off the coast of Iceland. Most of the matches to the breeding grounds in the Caribbean have been to the Silver Bank off Dominican Republic.

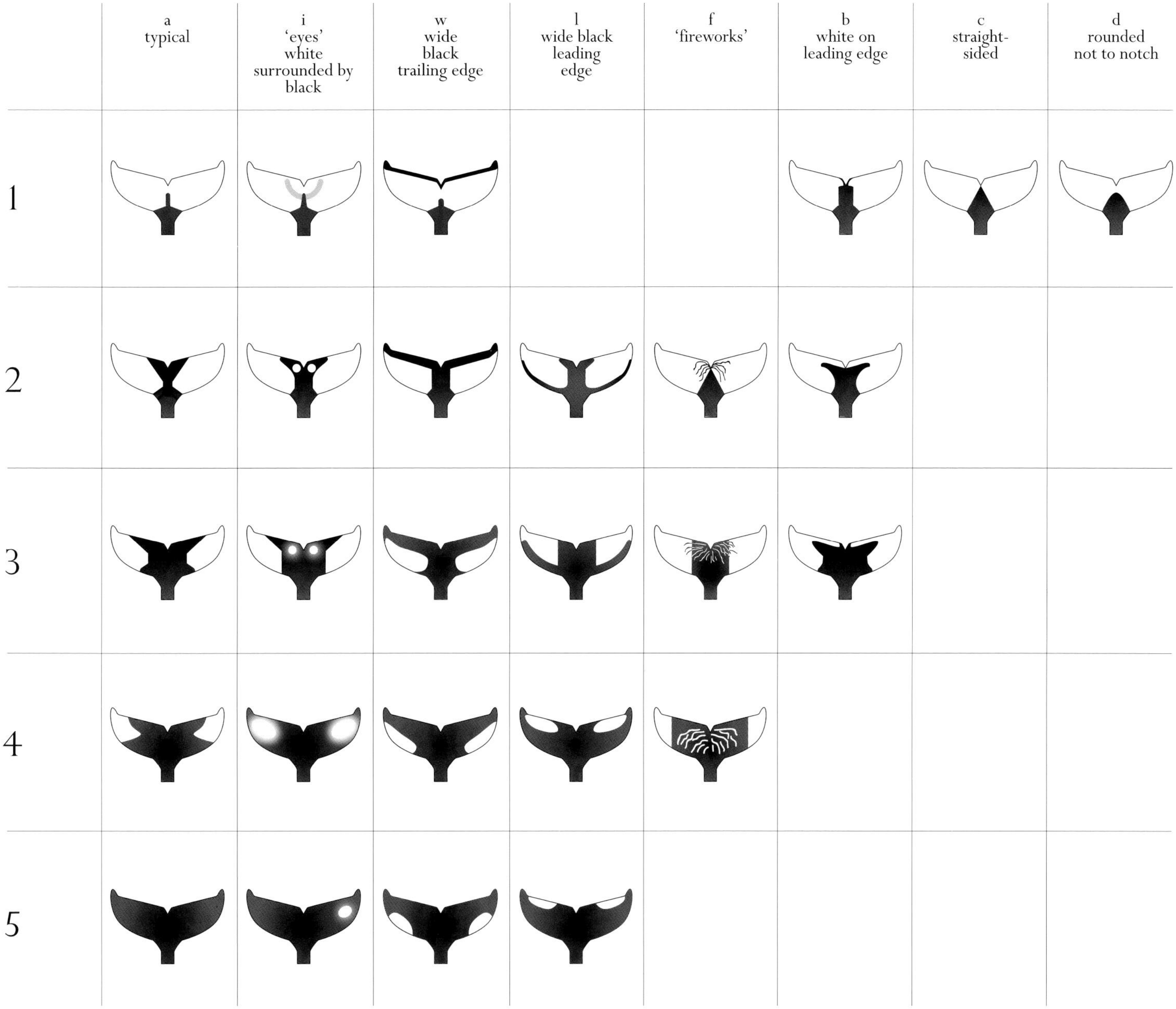

	a typical	i 'eyes' white surrounded by black	w wide black trailing edge	l wide black leading edge	f 'fireworks'	b white on leading edge	c straight- sided	d rounded not to notch
1								
2								
3								
4								
5								

ACKNOWLEDGEMENTS

Over the past four years that I have dedicated to pursuing my passion for the humpbacks I have been helped by the generosity of numerous individuals. Thanks in particular to Jim and Debbie Butterfield who made the very first donation and have continued to contribute to our work on the North Atlantic humpbacks through the different phases of the research, the film and now with this book. None of this could have happened without their generous support. Thanks also to the Atlantic Conservation Partnership for continuing to fund my research on the humpbacks.

Beyond the library of scientific papers that have contributed to my knowledge of the humpbacks, there are many whale enthusiasts who have generously educated me in the ways of the whales. Over the years, many of them have become friends. Flip Nicklin hospitably invited me to Hawaii to stay with him and generously shared his extensive experience with the humpbacks accumulated over the decades. Dr Jim Darling also kindly invited me out on their project boat in Maui, shared his insights into whale song and reviewed an early version of the manuscript. Dr Steven Katona, Judy Allen, Dr Peter Stevick and Rosemary Seton at the research group Allied Whale at College of the Atlantic, Bar Harbor, Maine, helpfully shared their years of information and data. Dr Phil Clapham and Dr Jooke Robbins frequently answered my numerous questions within minutes of sending them an email, no matter where they were. I have to wonder when they sleep. During my summer visits to Halifax, Nova Scotia, Dr Hal Whitehead of Dalhousie University was generous with his time and extensive understanding of whales. Drawing on his many years researching whales in Bermuda, Dr Greg Stone provided his wisdom and has since become a friend and neighbour in New Zealand. Dr Nils Øien, of the Marine Mammals Research Group, Institute of Marine Research, Bergen, Norway, provided beautiful fluke photographs from Norwegian waters. Tenna Kragh Boye at the University of Aarhus in Denmark shared her collection of humpback flukes from Greenland. Dr Rebecca Dunlop, Cetacean Ecology and Acoustics Laboratory, University of Queensland, Australia, provided her discerning feedback on the manuscript. Roger Etcheberry from St Pierre and Miquelon deserves special mention for his contributions to our North Atlantic whale fluke catalogues and his peerless matching capability. Reg Kempen, out of Trinity, Newfoundland, also provided a wealth of fluke IDs from his area. PhD candidates Tara Stevens in Newfoundland and Hilary Moore in Nova Scotia both offered all-embracing comments on the manuscript and I'm indebted to them for their thoroughness. Any mistakes or errors that remain in the manuscript are of course my own. Special thanks to The Secretary of State of the Environment and Natural Resources for permitting me to film humpback whale sequences on the Silver Bank, part of the Sanctuary of the Marine Mammals of the Dominican Republic. Thanks also to Michael J. Murphy.

There are many friends and family who have followed my lead to become whale enthusiasts. I couldn't have achieved what I have without the dedication of this handful of family and friends who have become volunteers and given thousands of hours of their time to help find the humpback whales, photograph them and catalogue their fluke IDs. This book is the culmination of their committed teamwork. In particular, Camilla Stringer has been a steady and stalwart volunteer from the early days when just the two of us started looking for our elusive whales in a small single-engine boat. Her background as a librarian and teacher has added a level of professionalism to our data collection and our Bermuda fluke ID catalogue would not exist without her careful administration. Michael Smith not only captained his comfortable boat, *Sea Slipper*, he is also devoted to being on the water despite the long hours, sometimes in adverse weather conditions. Michael's zeal is matched by his safety-mindedness when I am in the water 15 miles (24 km) offshore from Bermuda in the middle of the ocean with two young children and a wife waiting for me at home. Many thanks to Roland Lines who stepped in during 2011. Judie Clee, too, has backstopped our field (ocean) research and she gives us all a sense of teamwork by providing countless hours organizing and matching flukes in our Bermuda fluke ID catalogue. Bob Steinhoff also provided his boat to look for whales and has supported this project from its inception in many ways. My sister Jackie has stepped in whenever needed. Carol Dixon has been the behind-the-scenes webmistress, providing technical expertise on our website, www.whalesbermuda.com.

Thanks also to Peter Duncan, Managing Director at Constable and Robinson, for suggesting that I publish this book and to Duncan Proudfoot, Commissioning Editor, for gently guiding the process from beginning to end.

A special thanks to my wife Annabel, who encourages me to pursue my passions and covers for me at home during long days when I am on the water. And finally, a debt of gratitude goes to my two daughters, Elsa and Somers, who inspire me to work with the whales while keeping it all in perspective. Let's hope the world's oceans are in better shape when you are old enough to pursue your dreams.

Andrew Stevenson,
Bermuda

Additional photographs by:

Annabel Carter: pp. 6, 44 (bottom left), 82, 46; Kevin Horsfield: pp. 14, 15, 16, 17, 18; Roland Lines: p. 32 (top left); Camilla Stringer: pp. 32 (bottom), 136 (bottom left), 143 (bottom right); Michael Smith: pp. 34 (bottom), 77, 128 (bottom left), 136 (bottom right); Dr James Wood: p. 38 (top four photographs); Dr Hal Whitehead: p. 44 (top middle); Bob Steinhoff: p. 69; Reg Kempen: p. 70 (top); Judie Clee: p. 78 (bottom left); Jackie Stevenson: p. 79 (middle right); Dominick Lemarie: pp. 92 (top left), 96 (bottom right); Dan Henriksen: p. 95 (bottom left); Richard Lee: p. 141 (bottom left); Elsa Stevenson: p. 151 (top and bottom).